Farmers Must Be Co-operators

by C.O. Drayton

with an introduction by Roger Chambers

This work contains material that was originally published in 1914.

This publication was created and published for the public benefit, utilizing public funding and is within the Public Domain.

This edition is reprinted for educational purposes and in accordance with all applicable Federal Laws.

Introduction Copyright 2018 by Roger Chambers

Self Reliance Books

Get more historic titles on animal and stock breeding, gardening and old fashioned skills by visiting us at:

http://selfreliancebooks.blogspot.com/

Introduction

I am pleased to present yet another title in our "How To ..." series.

The work is in the Public Domain and is re-printed here in accordance with Federal Laws.

As with all reprinted books of this age that are intended to perfectly reproduce the original edition, considerable pains and effort had to be undertaken to correct fading and sometimes outright damage to existing proofs of this title. At times, this task is quite monumental, requiring an almost total "rebuilding" of some pages from digital proofs of multiple copies. Despite this, imperfections still sometimes exist in the final proof and may detract from the visual appearance of the text.

I hope you enjoy reading this book as much as I enjoyed making it available to readers again.

Roger Chambers

With Malice toward None and Charity for All

I dedicate this book to all Golden Rule Co-operaton.

Yours for Co-operation
C. O. DRAYTON.

A NATIONAL UNION OF FARMERS.

The foundation of a Farmers' Union is a strong, live local Union at all good shipping points; and often they are a success in the country school houses. All classes of farmers should be united in the local Unions, whatever the crop or crops may be which they produce. Every farmer should be induced to join a local Union, and the fact that he is a regular member should make his entire family faithful members. We wish to include the women and children in this great fight for economic freedom. The local Union should be a regular farmers' club, held the first Saturday of each month, in day time in the winter and at night in the summer. This Saturday must be regarded as Farmers' Union day, and every member must feel under obligations to quit work and take his family to the meetings.

Every regular member must pay one dollar a year national dues, this must be paid for him by his Equity Exchange if he is a stockholder. A good commodious hall should be built for a meeting place. When there is a strong Union, build a large store with a commodious hall over it and have a home for your local Union. The social fraternal feature must not be overlooked nor neglected in this great movement. We must use every legitimate power to bring the farmers into a more friendly relation, and make them more fraternal. They are too isolated, too separated, too selfish. This is there great weakness.

A good co-operative paper must reach every members' home weekly, teaching him golden rule co-operation, reminding him of the meeting suggesting topics for discussion and ways of increasing the power, efficiency and interest of the Union and of business requiring special attention at that time.

The president should appoint a committee on program for each month, who should see that there is a program of music, declamations, readings, debates and speeches for each meeting.

The lecturer should be the chairman of the committee on program, and should see to it that teachers, editors, lecturers and other educators are invited to address the meetings. Corn and stock shows, debates and declamation contests should be held at times to keep up interest in the meetings. The time and attention of the neighborhood would thus be profitably occupied, and at the same time a strong business organization be sustained worth thousands of dollars to the community each year. The farmers must be organized and educated to attend to their own business. Our pass word is, "Mind your own business." We must become co-operators in business. As soon as one hundred or more farmers are united around a good shipping point, steps should be taken to organize an Equity Exchange at that place. We mean that a stock company should be organized among the members. Only members must be allowed to take stock in an Equity Exchange. Put the shares at twenty five dollars each, and the limit to four shares. The Equity Exchange must be chartered by the National Union and by the state in which it operates. An elevator, warehouse and coal shed necessary for the business must be erected. All farm produce including live stock, cream, wool, eggs and poultry must be bought by this Exchange. Coal, flour, feed, salt, cement, fertilizer, fencing, twine and all farm machinery must be sold. All must be bought and sold on a safe margin. Out of the gross earnings must be paid the running expenses, interest on capital borrowed, each members' national dues, and not over 5 per cent dividends on the stock subscribed.

All earnings over this shall be net earnings. The net earnings are prorated back to the members who are stockholders according to the patronage they have given. If the net earnings are 10 per cent of the business transacted during the year, then each member who had bought and sold to the extent of five hundred dollars in one year would have prorated to him fifty dollars in cash as his share of the net earnings. He earned it, not by having one hundred dollars invested in the elevator but by his patronage. Those who have partonized the elevator or exchange to the extent of one thousand dollars would have one hundred dollars in cash returned to them— ten per cent of the business they have furnished being paid back for their patronage. If a member has only one share he would be paid in shares till he has the limit—four shares—when he would receive cash payments. Non-members should be paid for their produce the same price as members, and sell to them as cheaply, but at the end of the year prorate all of the net earnings among the stockholders. In this way you make a difference between members and non-members, and induce them to come in and stick. If a farmer finds from year to year that his neighbors who are members get back from twenty to fifty dollars for their patronage, he will finally wake up and go and hunt up the secretary and join the Union. We will have to show them that they are losing money right at home by not being members of the Union. Then they will come in and stick and many will not come before.

Give no member back anything until he has four shares—the limit. Give him shares instead and increase your capital. We have at the present writing nearly one hundred Exchanges where two hundred farmers are united in the Equity Union, each with one hundred dollars invested and all true co-operators. Each member knows that his Union handles his produce without profit. Every cent made on what he buys or sells is prorated back to him at the end of the year. Each farmer on the outside soon sees that he is losing money by not being a member and a stockholder, as members get a benefit which he cannot get until he becomes a member. This Union works for benefits for its members only. The stubborn farmer will finally have it rubbed into him so hard that "it pays to co-operate" that he will finally come in. This will be the entering wedge in every community that will finally destroy the profit system, which is a robber system. A system which enables thousands of greedy, selfish millionaires to combine their wealth in such a way as to continually threaten the masses with industrial slavery. A good strong live local Union, holding one meeting each month, with an interesting literary and musical program, incorporated and chartered as a corporative business organization, are the corner stones, the real foundation of the great Industrial Union we have in mind and are giving our life to erect. Each of these local Unions will be a link in the great National Union chain. They must be bound together under one national head. The National Union must be composed of its National Officers, including a Board of Directors and of representatives of all the local Unions. This National Union must be the Supreme Head and liberally and loyally supported by every Exchange and every member. Under this one head must be united cotton, grain, wool, stock, fruit and produce growers. The National Union must be a great organizing force. There must be a continual campaign of organization, and education among the millions of farmers of our country. Money must be spent freely for this purpose. A good co-operators' paper must reach every member weekly, which teaches co-operative marketing as well as scientific production. Hundreds of organizers and educators must be kept in the field agitating, educating, organizing and pursuading the farmers to become golden rule co-operators in business. The educational feature in this movement, so necessary to success must not be overlooked nor under estimated. The members must be continually stirred, aroused, interested, educated. Every local Union must be kept busy and growing. The efforts of the National Union should be concentrated on the organization and building up of local Unions, Equity Exchanges and Creamery and Mercantile Exchanges.

THE FARMERS' EQUITY UNION

The Farmers' Equity Union is a National Union of farmers with headquarters at Greenville, Illinois which is now started in Ohio, Indiana, Illinois, Missouri, Oklahoma, Kansas, Colorado, Nebraska, South Dakota, North Dakota and Minnesota. The President is C. O. Drayton, Greenville, Illinois, Vice-president, R. Romer, Liberal, Kansas, Secretary, Geo. L. Denny, Greenville, Illinois and the treasurer, State Bank of Hoiles & Sons, Greenville, Illinois. The other directors are Edwin W. Reed, Lux, Nebraska, E. L. Waters, Yuma, Colorado, S. S. Ray, Cyrene, Missouri, T. L. Line, Columbia City, Indiana, A. Hoffman, Leola, South Dakota, R. L. Cook, Guymon, Oklahoma, Chas. Kraft, Odessa, Minnesota and F. J. Taylor, Lucas, Ohio.

Section 2, Article 1 of the constitution says: "The objects of this

Union are to promote intelligence, morality, sociability and fraternalism among its members, and to secure fair dealing in all the business relations of farm and mercantile life." Co-operation to the advantage of all of our members is our chief object. Our work is strictly and purely educational, but results in successful co-operative business organizations wherever we are organized.

Strong local Unions are being organized at the best shipping points in the states where we have started. An Exchange is established as soon as practicable. This is a stock company for co-operative buying and selling. The shares are twenty five dollars each. The limit, four shares to each member. Only members can be stockholders. These Exchanges handle grain, hay, stock, cotton, and all farm produce. Also flour, feed, coal, salt, fertilizer, twine, wagons, lumber, and all farm machinery. We buy and sell on a safe margin the same as the markets around us. We never boost prices on what we buy nor cut prices on what we sell. Then no assessments on members will be necessary. At the end of each year we are sure to have a gross earnings, if we get the patronage of our members. Out of this gross earnings the Board of Directors are authorized to take running expenses, necessary repairs, each members' national dues and not exceeding 5 per cent dividends on the stock subscribed. All net earnings are prorated back to stockholders according to their patronage. We are uniting the trade at many of our Equity Exchanges and doing more than two hundred thousand dollars worth of business each year. We are paying back at a number of places from three to ten thousand dollars for patronage. This plan brings a large volume of trade to one center and reduces the cost of handling. Our partonage dividends run from three to twelve per cent. Mott, North Dakota owns two large elevators, two lumber yards and has a membership of nearly three hundred good farmers, and in two years prorated back nearly fifteen thousand dollars for patronage. This large sum of money was scattered out among the farmers and their families instead of being concentrated in the hands of a few rich speculators in Minneapolis. The farmers at each town where we have an Equity Union soon learn that it pays to be a member of the Equity Union. The plan bids for members, stockholders and patronage. It will unite one milion farmers in the Equity Union in a few years. It brings a large volume of trade to one center and thus decreases the cost of handling. Concentration and co-operation are the watch-words of the business world. Millions of farmers will be educated through the Equity Union to follow the same up-to-date principles.

Equity Union is starting the wedge that will split asunder the profit system which now robs the people of billions of dollars annually. It will bring a fair distribution of wealth to the producers and work no injustice to consumers.

Under this co-operative system we will have more home owners and builders and not so many millionaires. The wealth produced in the country will accumulate there and be used to build roads, schools, churches and beautiful country homes, and more of our boys and girls will stay on the farms.

The foundation of success in this grand movement is strong local Unions at every good shipping point. But the local and national Unions should unite in a campaign of organization and education that will result in the establishment of an Equity Exchange and educate the farmers to center their trade with that Exchange.

One great work of the Equity Union is to unite the two thousand elevator companies under one National Head, persuade them to adopt true golden rule co-operation and induce a large majority of farmers at each market to unite their patronage in the meantime organizing new Exchanges as rapidly as possible. We will then have an organizing educating force that will result in three thousand more Exchanges in afew years. This will mean five thousand Exchanges united under one National Head, with an average membership of two hundred farmers or a million farmers united. This will mean terminal elevators, Equity Union mills, warehouses, stock yards and large wholesale plants in the cities for the distribution for dairy products, eggs, poultry, fruit, and all farm produce, all owned and controlled by producers and consumers instead of the speculators and trusts.

A MILLION FARMERS UNITED

The possibilities of one milion farmers united in a fraternal co-operative, industrial Union are beyond comprehension or calculation. We must unite under one directing center. Man has a brain center with nerves extending to every part of the body. This is a good model for united action.

The plan of the Farmers' Equity Union is to unite one million farmers and educators and consumers under one Supreme Head. The work of this Supreme Head is to carry on a continual campaign of organization and education among the farmers. It must be the great organizing, educating force which is absolutely necessary to success in this great movement. It must enlist in its service the very best organizers and educators and keep them constantly in the field agitating and educating. This National Union must connect each member with all the others. Each member will be for all the rest and all the rest for him. This will be the most powerful and beneficial fraternal organization in our country. Including the farmers' families, which should by all means be done, it will number five million members in its ranks.

The power of the individual farmer in such a Union can only be imagined. He no longer fights his battle single handed with powerful combination in the business world, but instead he will have the power of one milion farmers behind him, preventing a low unjust price on all he has to sell and a high extortionate price on all he buys. There will be wonderful protection for each member in such a Union.

In November of each year this National Union of one million farmers will contract with a reliable cordage company for one hundred million or more pounds of twine for its members and others for the following year.

We will be able to take the entire output of one factory for one year. We put that factory in a position to give us the very best terms possible. We remove them from the competitive field and take them over into our co-operative world where they are protected and benefited as well as we.

When organized under one head we will center our entire trade with one reliable factory for wagons with another for gasoline engines and all farm machinery, with another for automobiles, with another for harness, furniture, cement, fertilizer, etc.

Reliable factories will bid for our large cash patronage when it is organized. Our five thousand Equity Exchanges, at the best shipping points each capitalized at from ten to twenty thousand dollars will be behind the proposition to give it financial standing and support, and will be our ware-housing and distributing agencies. Members will have a benefit here that will bring outsiders in by the thousands and tens of thousands. When this plan is carried out it will surely unite two million farmers instead of one. Independent factories will spring up and be sustained by this movement and thus thwart the scheme of the trust that seeks to monopolize the business and rob the people.

We will center our patronage through our own directing National Union and one dollar per annum from each member will be fully sufficient for all expenses.

Brother farmer, this is not only a dream but will be made an actual fact by the Farmers' Equity Union. You know it can be done. The only thing holding it back is the farmers on the outside who refuse to join the Union and support it with eight and one third cents per month. Are you one of these farmers who is "wearing out the holdback straps?"

HOW TO UNITE THEM

But how are we to get one million farmers and their families united? This is the great problem which the Farmers' Equity Union undertakes to solve and is solving successfully in eleven states where we are started. This Union must be grown. It will grow as honest, sincere, able lecturers are put into the field and kept steadily at work. The Equity Union Exchange, our official paper must be doubled in circulation every year. The encouraging news each month from our Unions and Exchanges which is reported in this paper and the many fine articles on golden rule co-operation all educate and instruct our members and build up our Union.

Every Farmers' Elevator Company ought to resolve itself into a local Union of the Farmers' Equity Union and adopt our Exchange By-laws which embody our plan and principles of golden rule co-operation. Our greatest danger is that we as farmers, will under estimate the power and influence and imperative necessity of a continual campaign of organization and education to unite and keep united one million farmers.

The next important step is that the members at each shipping point should organize an Equity Exchange and have their own elevator, coal shed, warehouses and stock yards. Very few local Unions will live unless organized for business.

This stock company ought to be chartered by the state in which it operates and must be on the strictly co-operative plan. The shares must be twenty five dollars each. And the limit four shares. Only Equity Union members must be allowed to take stock. This Exchange must handle grain, hay, stock, eggs, poultry, coal, tile, cement, flour, salt, twine, farm machinery and automobiles. Buy and sell on a safe margin. Out of the gross earnings take running expenses, national dues and not exceeding 5 per cent dividends on the stock subscribed. Be sure to put it in the by-laws of your Exchange that the directors cannot declare a dividend of over five per cent on the stock subscribed. A three per cent stock dividend would be still better.

After the elements of cost are all taken out then all over this is net earnings. All net earnings are prorated back to stockholders according to the amount of patronage given. If you have five thousand dollars net earnings made on a business of one hundred thousand dolars furnished by the stockholders you have five per cent patronage dividend to prorate either in cash or shares. Each stockholder would receive five per cent of the patronage he gave the company during the year. Five hundred dollars patronage would bring twenty five dollars in cash or another share back to the patron if a stockholder. If he had only one share, then, instead of cash, he would be given another share and the twenty five dollars added to the capital of the company. Nothing should be paid back to any stockholder till he has four shares, the limit. This means a ten thousand dollar Union wherever we have one hundred stockholders. This ten thousand dollars cannot be prorated back but will be invested in a good grain elevator, warehouses and coal sheds and furnish some capital for business. We pay outsiders as much for their produce as stockholders. We sell to them as cheaply as to our members, but we do not prorate back to them any cash or shares till they come in and become members and stockholders. We can lick every trust in the land if we get enough farmers united, and the Equity Union is uniting them.

The National and Local Unions must unite in a continual effort for trade at the Farmers' Elevator or Equity Exchange. The larger volume of trade we center together, the more economy in handling and the more probability of prorating cash for patronage.

At many of our Exchanges the farmers on the outside find that members are receiving from year to year a patronage dividend of from ten to one hundred dollars and hence they are hunting up the secretary, joining the Union and taking stock in the Exchange. Their own selfish interests are driving them into the Union.

Wherever our lecturers and members can say to the farmers on the outside, take stock, hall your grain to our elevators and you will receive back in cash all profit over five per cent on the money invested, it brings patronage so necessary to success. Our plan leads to centralizing our trade. At many good towns in the west we find from three to seven grain elevators owned by line companies who are there to rob each community of thousands of dollars. By concentrating our trade with one elevator in an Equity Exchange, we have the biggest and best business in the town and all profits are paid to the farmers instead of to unnecessary middle men.

Brother farmers, our power and best interests in business is in uniting our trade and organizing our patronage nationally. Let us no longer be narrow minded nor short sighted but broad minded and generous hearted.

When enough farmers are united and educated on the Equity Union plan we will have our own terminal elevators in the large central markets, our stock yards outside the cities, and finally our central plants or markets in the cities to receive and distribute dairy products, fruits, vegetables, flour, meat products and all produce direct to consumers' kitchens.

We must put Equity Exchanges at five thousand of the best shipping points. Nearly two thousand towns have Farmers' Elevator Companies. The most of them are capitalistic in plan and principle and are therefore owned by a few men. They must be organized on our golden rule plan of co-operation which will take in ninety per cent of the best farmers at each town. These Exchanges then will all be under our one National Head. This work of organization must be pushed by the National Union, the Local Union and every member must push the work of organization till one million farmers are solidly united. We now have over one hundred live, active, growing Unions, some of which have nearly three hundred members. We have made a fine start in three years by planting this Union in eleven states and have every reason to be encouraged and sure of ultimate success.

CO-OPERATION VS. THE PROFIT SYSTEM

True blue golden rule co-operation vs. the profit system is an issue in the economic world which we are bringing to the front more and more every day by this grand movement. This movement is so important to the millions of farmers who produce every year billions of dollars of wealth that the weekly paper which teaches farmers the true principles of co-operation deserves a million circulation. The capitalistic profit system is the great robber system of our day. It drains the country districts of untold wealth each year and centers it in the hands of the few, selfish rich, who combine their millions of wealth and use them to fasten on the wealth producers a system of industrial slavery. The profit system makes the rich richer and the poor poorer. Golden rule co-operation will give the wealth producer all he produces. In the profit system the dollar is the unit and the idol. In the co-operative system the man is the unit. Co-operation enlarges our humanity and makes us more fraternal. We have had co-operation of the few to the sorrow of the many for decades. We must educate the millions of farmers and their families to be co-operators. They are human beings with brains and ordinary sense, and can be educated.

If every agricultural paper would give two columns each issue to this subject and urge the farmers to unite under one National Head in the Farmers' Equity Union, we would have our million members in a few years and demonstrate our principles of true blue golden rule co-operation so fully that the farmers would never be separated and robbed again. Farmers would learn that combination, concentration and co-operation bring power, protection and prosperity in the business world. We ask the agricultural papers to investigate our National Head and the operations of our successful Exchanges and they will be convinced of our honesty and sincerity and that our plan is sane, sensible and successful and that our movement is worthy of the approval and sanction of every paper which is the friend of the farmers.

We as farmers must overcome our selfishness and distrust of one another and learn that unless we hang together we will hang separately. Our selfishness and distrust separate us and weaken us and make us an easy prey in the commercial world for our enemies who continually rob us. We are an easy mark as long as we stand as individuals. The overcoming of our selfishness and narrowness as farmers will make us better to our families, better neighbors, better citizens and more powerful in the political and commercial world. The seven million farmers of the United States can and will rule this country in commerce and politics when united. We can easily put out of business every speculator, profit taker, grafter and extortioner and break the power of every trust to rob the people when we are thoroughly united. We must come together as farmers in the Farmers' Equity Union with a full determination to do right by each other. We must be broad minded and generous hearted. Each must be for all the others and all the others must be for him. Fraternalism is the twin sister of co-operation.

The object of the Farmers' Equity Union is to promote intelligence, morality, sociability and fraternalism among its members and to educate them to be true co-operators. It is an organizing, educating force. More than one hundred strong local Unions are already organized at good shipping points, and an Equity Exchange is established and demonstrating true blue co-operation. This is a stock company for co-operative buying and selling. The shares are twenty five dollars each. The limit, four shares to each member. Only members can be stockholders. These Exchanges handle grain, hay, stock, cotton, wool, lambs, calves, eggs, poultry, dairy products and all farm produce; also flour, feed, coal, fencing, salt, wagons, autos and all farm machinery. We buy and sell on a safe margin the same as the markets around us. Then no assessments for losses will be necessary. At the end of each year we have a gross earnings. Out of this gross earnings the Board of Directors are authorized to take running expenses, necessary repairs and five per cent dividends on the stock. All over this is net earnings or profit. The directors are prohibited from declaring over five per cent dividends on the stock subscribed. All net earnings are prorated back to stockholders acording to their patronage. This is true co-operation. Each members' produce is handled at cost, without profit. School house campaigns are made in every state where we are started, organizing Exchanges and educating the farmers away from the profit system to be true blue co-operators. We are uniting the trade of from two to three hundred farmers at many of our Equity Exchanges and doing a business of from two to three hundred thousand dollars each year. We are educating the farmers to center their trade with their own Equity Exchanges by prorating in patronage dividends of from three to ten thousand dollars each year. We are handling each members' trade at cost without profit. We pay him as much as he can get at any other market and pay back to him in cash all profits. If we had made five per cent on the business then every share holder who gave one thousand dollars of trade during the year receives fifty dollars back for his patronage. Not because he had one hundred dollars of the stock. That is his part of the working capital, but because he gave us one thousand dollars worth of his patronage. **This is true co-operation.** It will bring a large volume of trade to one center and reduce the cost of handling to the minimum.

Each share holder who gives us five hundred dollars worth of patronage during the year would receive twenty five dollars for his partonage if he holds four shares—the limit. If he holds less than four shares, we give him another share and add the twenty five dollars to our capital. We build up his shares and our capital till he has the limit—four shares, and then we prorate to him in cash. We will finally have two hundred farmers united, with exactly one hundred dollars each in the Exchange, and with a capital of twenty thousand dollars invested and banked. We will pay non-members as much as members, and sell to them as cheaply but we will not prorate to them anything until they become members and stockholders.

If a farmer finds that his Equity Union neighbors receive just as much as he does for their produce and buy as cheaply, but at the end of each year divide up a five or ten thousand dollar melon, he will finally be induced to join. The plan bids for members, stockholders and patronage. It will unite the farmers and keep them united. Combination, Concentration and Co-operation are the three C's that will bring Power, Protection and Prosperity to the millions of farmers—the three P's. These are the watch words in the business world. Millions of farmers will be educated through Equity Union to follow the same up-to-date principles. Equity Union is the entering wedge that will split asunder the profit system which now robs the people of billions of dollars every year. It will bring a fair, honest distribution of wealth to the producers and work no injustice to consumers. Under this co-operative system we will have more home owners and builders and not so many millionaires. The wealth produced in the country will accumulate there and be used to build roads, schools, churches and beautiful country homes.

THE VALUE OF THE FARMERS' PATRONAGE

The Farmers' Equity Union is teaching the farmers the value of their patronage. The yearly patronage of the three hundred farmers around Greenville, Illinois will amount to more than three hundred thousand dollars on milk, hay, stock, wool, eggs, poultry, flour, feed, fencing, twine, salt, wagons and all farm machinery. More than five per cent clear profit is made on this patronage. Five per cent is a very conservative estimate. So the patronage of these three hundred farmers is worth at least five per cent of three hundred thousand dollars, or fifteen thousand dollars annually. These three hundred farmers are earning this large sum of money each year by their patronage, and then giving it away to unnecessary middle men.

The one thousand Equity links we are working hard to establish at the best shipping points will be worth on the average, ten thousand dollars each to the members every year, or ten million dollars. Brother farmer, will it not pay to put a few dollars into an industrial Union that will so organize our patronage so as to put the ten million dollars back into the pockets of the farmers who worked hard to earn it, instead of giving it away to speculators and unnecessary middle men?

The average farmers' patronage is worth about fifty dollars per year. Equity Union will put that fifty dollars back into the farmers' pocket each year. So the patronage of the three million renters on our farms is worth one hundred fifty million dollars every year. They are earning this vast sum each year by their patronage and then giving it away to unnecessary middle men. Equity Union would put every dollar back in their pockets to whom it justly belongs.

The two million farmers who have mortgages are losing one hundred million dollars each year by giving away their patronage. These two classes of plow handle farmers, who need every dollar they can get for the common comforts of life and for the purchase of homes, are truly and surely sacrificing two hundred and fifty million dollars worth of patronage year after year.

The farmers around every good shipping point are losing enough every year to build and equip a good elevator and all necessary ware houses, by not organizing and centering their trade in one co-operative channel.

The Equity Union idea is for the three hundred patrons to each put in one hundred dollars as his share of the working capital, buy and sell on a safe margin, and at the end of each year pay back to members all net earnings. Prorate back to members according to their patronage, or in other words, pay them what they have earned by their patronage. If the net earnings are all prorated back for patronage, then only patrons will hold the stock, and the business is run by and for the patrons. The patrons run their own business instead of Mr. Capitalists. This is Equity Union, and gives the wealth producer all he produces. Mr. Capitalists has no chance to rob him. Give us one thousand Equity Exchanges, centering their trade together for farm machinery with individual factories, and we can reduce the price to our members from thirty to fifty per cent. We come to each factory with our patronage organized. We save expense of advertising and traveling salesmen, and take away all risk of sale. We take the factory out of the competitive world with all its worry, uncertainty and expense into our co-operative world. We only pay for labor, good material, and a very small profit on each wagon or machine. The same wagons which now sell for seventy five dollars can be furnished to our members for fifty dollars if we have enough members buying together. Self-binders which now cost one hundred fifty dollars can be sold to members for seventy five dollars. Brother farmer are you still wearing out the hold back straps on this proposition and holding back our grand movement which is trying so hard to give you farm machinery at half price? The generous hearted, public spirited farmers are uniting and helping us to work out our plan of co-operation.

What a help this movement will be to our three million renters who must furnish all of the farm machinery for three million farms. Brother farmer, go out to the machine shed with your pencil and note book and figure out how much you paid for all of that machinery when new, add it all up and divide the sum total by two, and you will see what you have lost by not having an Equity Union at your town. Call a meeting of farmers at once, read these articles about the Equity Union, discuss the matter fully and arrange a big meeting. Write to the Equity Union, Greenville, Illinois for a speaker and we will organize a Union at your town with over one hundred members.

There is one thing sure, we are robbing ourselves of millions of dollars every year by giving away our patronage. A want of organization, education and co-operation is the cause. We must be organized on the Equity Union plan—a truly co-operative plan which gives the producer all of the wealth he produces. Farmers must become willing to quit work and go to an Equity Union meeting the first Saturday afternoon of every month. We ask them to give twelve half days or only six whole days of each year to Equity Union meetings. Then they must be willing to pay one dollar per year dues to support a Union. A continual campaign of organization and education is absolutely necessary to success. This cannot be done without meetings, money and members.

When one million farmers meet regularly once every month, pay one dollar per year national dues, and all read the same weekly co-operative paper, which teaches scientific production and co-operative buying and selling, there will be a complete revolution in the farmers' world, the business world and the political world.

A NATIONAL INDUSTRIAL UNION NOT A POLITICAL PARTY

The importance and power of a National Union of farmers cannot be emphasized too much. It is often said that Grant and Lincoln saved our country, but neither of them could have accomplished what they did if there had not been an organized force behind them which gave them means, instruments and power to suppress the rebellion. Organization is the first requisite when any great work is to be accomplished by the people. Organization saved our country. Colonel Roosevelt has repeatedly urged the farmers to organize for the betterment of their conditions. Ex-secretary of agriculture Wilson advises us to organize a system of co-operative buying and selling, showing in his report that we as producers receive only about half of what the consumers pay. The rest is wasted in distribution or accumulates in the hands of millionaires. When we the people, are really Uncle Sam, our government will use part of the millions of dollars now spent to educate us to produce, to teach us how to buy and sell co-operatively. The great need of this education is very apparent when we consider the fact, shown in secretary Wilson's year book for 1907, that we produced one hundred twenty million bushels more grain in 1906 than in 1905 and sold it for forty million dollars less money. In other words, we played into the hands of speculators by our competitive system of marketing in such a way as to lose the one hundred and twenty million bushels of grain and forty million dollars in cash besides in low unjust prices.

Ex-president Taft said that the overthrow of socialism is our greatest problem. But we would rather say that the wise solution of our economic problems is one of the greatest questions now before the American people. The agitation by socialists against monoply, concentration of wealth and industrial slavery, and in favor of co-operation, combination of the people, fair distribution of wealth, equal rights and equal opportunites, are all educating in the right direction. All good ideas and principles evolved by socialism will be accepted and adopted by the American people. The American people do not want fraternalism. I use the term in the sense of doing for others what they should be educated and encouraged to do for themselves. The truly wise father will not long do for his children what they can and could learn to do for themselves. Uncle Sam will never teach his children to be self reliant and prosperous by government ownership of all property. If the government owned all of the coal mines, the conflict between labor and capital would simply be transferred from the coal barons to the government. If however, the half million miners of our country were thoroughly organized in a great industrial Union and educated to be truly fraternal and true co-operators, they would finally be induced to stop depositing their money in banks where it never comes out again, and each and every one of them would take shares in the mine. Then the miners would own the mines, or more than fifty per cent of them, and be both laborers and capitalists and conflict would cease. The national government ought to own forty per cent of the capital and the miners sixty per cent. The capital should never draw over two per cent and all earnings of the mine over this should be prorated to the miners who do the work.

Co-operative ownership will arouse in each individual a sense of manhood and responsibility conducive to the development of the highest type of citizenship. We are entirely opposed to any economic system that will destroy the individuality of the citizen. We believe that golden rule co-operation is possible without this.

If the two million employees of the railroad each owned shares of railroad stock and collectively owned sixty per cent interests in the great thoroughfares in our country, strikes and the paralizing of the business would be a thing of the past.

If the millions of wage earners in our factories and foundries would use their Unions to educate their members to be more fraternal, intelligent and moral, and to be true golden rule co-operators, they would soon be both capitalists and wage-earners, and all possibility of strikes and conflict would be removed. If one million farmers will unite under one National Head in a Union having as its chief objects the promotion of intelligence, morality, fraternalism and co-operation among its members, we will not need fraternalism nor any other Ism to bring us economic freedom. The day is coming when the farmers will own and control co-operatively the channel direct to the consumers. They will have an intelligent, economic system of distribution of their valuable crops, which will insure the producer full values and consumers more reasonable prices. By this system graft, extortion and speculation will be reduced to a minumum.

They will also organize their patronage and center their trade with individual factories and save millions of dollars now spent for advertising and traveling men. Our hope is in true blue golden rule co-operation. Industrial co-operative Unions will bring industrial freedom, by developing the intelligence and manhood of each individual member. It is not the function of government to run our individual business for us, but to so regulate and control all industrial organizations, corporation and combinations that injustice to the general public may be prevented and equal rights and equal opportunity may be guaranteed to every individual.

Industrial Unions must be kept free from partisan politics. The old farm organizations were either broken up or badly crippled for life by allowing politicians to come in and run them. This new organization must keep off of the rocks upon which the old ones were wrecked.

Any industrial Unions which allows a political faction or party to dominate it, is doomed. Our past experience clearly demonstrates this to be a fact. Instead of allowing any factions to dominate our Unions we will dominate all political parties.

The two million employees owning and running the rail roads would have more political power than all of the present day magnates combined. They would not be partisans. They would be like Jay Gould, "Sometimes republicans and sometimes democrats, but always railroad men."

The sixty thousand coal miners of Illinois have more power in securing friendly legislation than the two hundred thousand prosperous farmers, notwithstanding the fact that a large majority of the miners own only a pick and shovel, while the farmers own all of the rich land of Illinois. The cause is easily guessed. The miners are organized, while the farmers have only a small beginning of an industrial union in this great agricultural state. But the miners so far have not allowed their Union to drift in partisan politics. As an organization they allow each member to vote as he pleases. They do not disturb his party relations. They antagonize no political party. They are non-partisans. But when the legislature is in session they wield the power of their industrial organization. Through this power the miners have made the Illinois legislature do their bidding.

A million farmers united, fifty thousand in each of twenty states, will secure all needed legislation in a few years. Future legislatures will have no millionaires as members, but be composed of representatives of industrial Unions, and the United States Congress will be the great national umpire, regulating and controlling the great industrial Unions and corporations, so that equal and exact justice shall be rendered to all.

PRODUCTION AND PRICE OF FARM CROPS

Strenuous efforts are being made and millions of dollars are spent annually to instruct and pursuade farmers to produce larger crops. The railroads are making commendable efforts by their corn and wheat specials. The catalog houses and the International Harvesting Company are spending milions of dollars to educate farmers to produce more. Vast sums of money are spent annually for agricultural colleges, experiment stations, farmers' institutes and agricultural papers, all teaches intensive farming or how to produce more.

The writer is in full sympathy with this great movement for more scientific production, and believes that we, as farmers, cannot put too much intelligence into our business. We insist that the farmer needs more education than the banker or the merchant. In the center of every agricultural township there should be a high school where all the common branches are thoroughly reviewed, and in addition, a four year's course suitable for country boys and girls. This course should include domestic science, chemistry of soils and fertilizers and a study of plant and animal life. We should cut out the dead languages and teach the live languages of nature.

But I wish here to issue a word of warning to my brother farmers. Is it a safe business proposition for the farmers under the present conditions, to produce a bumper crop of any king? Every leading market in our country is largely controlled by an organized force of buyers.

The attorney general of Missouri, in his charges against the beef trust, claims that Armour, Swift, and Morris not only control the National Packing Co. of Chicago, but they also control twenty six other packing plants which were formerly independent. The individual stock man is powerless before such a combination. Every time a good hog crop is produced, the price is low and unjust. Two years out of every five the farmers produce large crops of hogs, and lose money on every one of them; then there is a shortage, with high prices for three years, when a majority of farmers have none to sell. The cause is plain. The farmers are a mob—the buyers are thoroughly organized.

For the same reason, we sell our largest and best grain crops for less money than our small, inferior crops. Ex-secretary Wilson showed in his report that in 1906 we produced one hundred twenty million more bushels of grain than in 1905, and sold it for forty million dollars less money. Why should a big crop of fine quality bring the farmers less money than a small inferior crop? Why spend millions of dollars educating us to produce big crops if we must, on that account sell them below cost of production? We sold the 1906 crop of wheat at least one hundred milion dollars under a fair just price. There was no over production. The railroads made millions of dollars hauling it, the speculators handling it and the millers grinding it. But the farmers, the real producers, lost millions of dollars producing and marketing it.

The price was reduced from 98c to 68c in St. Louis in July 1906 in fifteen days time, not because there was too much wheat in the country, but because there was too much on our central markets at one time. The farmers put so much on the market when they threshed that over three hundred car loads each day rolled into St. Louis and about the same into Kansas City, Omaha and Chicago, and when the northwest farmers threshed, fifteen hundred carloads were rolled into Minneapolis each day.

Brother farmers listen! Every effort possible is being made to teach us to produce bumper crops, but suppose our fifty million acres of wheat in 1915 should really average five bushels more per acre than it did in 1914, and on June 1st, 1915 the government report should estimate the crop at nine hundred fifty million bushels, and the speculators have that report printed in flaring headlines in all of our papers, how would the farmers market that 1915 crop? Hundreds of thousands of them would hurry to sell before the price goes down, and another valuable crop will be sold at a heavy loss to the farmers, while the railroads, speculators, millers and trusts reap another rich harvest at the expense of the great mob of unorganized farmers. We would lose enough on one crop, by not being organized, to build a fine country elevator at every good shipping point where wheat is sold, and a terminal elevator in every good central market. What is the remedy? Not less intelligence in production, but more co-operation in marketing. Co-operative buying and selling must go hand in hand with scientific production. We must be organized and equipped to hold the surplus of every great crop in the country until the demand calls for it. We must regulate the supply to the demand. We must quit competing for the markets and be co-operators.

The Farmers' Equity Union is continually agitating for strong local Unions at all of our best shipping points with an Equity Exchange for co-operative buying and selling. More granaries must be built on the farms; and we must have a minumum price on all cotton, grain, hay and stock. Every manufacturer and every business man has a minimum price below which he cannot afford to sell and does not sell. Business men prevent a low unjust price by co-operation. Farmers must use the same potent weapon.

When a strong majority of the grain raisers are united under the Equity Union banner and all read the same weekly co-operative paper, there will be united action, and good wheat will never sell below the cost of production in any of our leading markets. What will we do with the surplus? We will fan it out and feed it to the chickens and stock and have no surplus. Why place the intelligent American farmer on the same commercial or industrial level with the peasants of Russians or peons of India by insisting we shall sell wheat at the Liverpool price? If one hundred million bushels surplus shipped to Europe reduces our price on the entire crop one hundred million dollars, we will fan it out, feed it to the chickens and stock and sell and sow only the best of our crop.

James J. Hill claims we will have no surplus in a few years. What will we do with the poor men who cannot hold their grain? We will give them the market when we thresh. We thresh the seven hundred million bushels of wheat from Texas to North Dakota in July, August,

September and October. Our price during these four months ought to be one dollar per bushel in our leading markets. In November and December it should be $1.05; in January and February, $1.10; in March and April, $1.15 and in May and June, $1.20. We ought to grade the price up enough to cover shrinkage, storage, insurance and interest. This plan will break up the dumping system. We will not get in a hurry to sell before it goes down. We will not sell on a falling market. We will act together when thoroughly organized and all reading the Equity Union Exchange weekly. The little cost of organization will be nothing compared with the benefits. When we thresh the market will demand about one-third of the crop.

The one hundred million people will eat about one-third of the crop while we thresh it. One-third of the wheat sellers can sell all of their crop when we thresh. If wheat sellers were organized the millers would be glad to co-operate with them for steady prices. If prices were graded on wheat the same as they are on hard coal from month to month, millers all over the country would fill up their reserves at threshing time and make a market for the wheat for all of the poor men who could not hold. Farmers would have more grainaries and country and terminal elevators for storage and all who were able would hold their wheat off the market and thus protect the poor man and themselves from the dumping system. Farmers, why should wheat be $1.60 per bushel in June in 1909 and down to 80c in July? Why should the price be steady at $1.25 per bushel all winter and then down to 75c a bushel in July.

Is there no way to protect the poor man from the gamblers' hold up? Is the wealthy farmer so selfish and short sighted that he will not lend his influence and pay money to build up a Union that will protect himself and every poor tenant from this hold up by the speculators every time we have a good crop? The price of grain, stock and all food products ought to be just as sure and steady as the price of coal, sugar, or farm machinery and will be when farmers unite and sell co-operatively.

The Farmers' Equity Union will teach intelligent production and co-operative marketing. Then we will not be afraid of bountiful crops to feed the people, knowing that each crop is all needed before the next is ready, or if there is a surplus, a short year is sure to come along and absorb it. Brother farmer, start a Union at your place. All great movements have a small beginning. Get three members as a nucleus, get a charter from Greenville, Illinois and we will come and assist you till you have one hundred members organized in an Equity Exchange.

TWO EQUITY IDEAS

The two principle Equity ideas are fundamental to the success of any and every farmers' union. Education and co-operation will unite the farmers and keep them united. Nothing else will. We are insisting throughout this text book upon the importance—the absolute necessity—of a continual campaign of organization and education. We are sure there can be no permanent success only as the farmers are continually stirred, aroused, informed and instructed. Eternal vigilance is the price of liberty. Ignorance, prejudice, narrowness and selfishness must be overcome. These things separate, weaken and enslave us, and make us an easy mark for speculators, gamblers, profit-takers and all combinations. Economic freedom, like political and religious liberty, are only possible to an intelligent, moral, fraternal and united people. He who would be free must himself strike the blow. How will we get the farmers to strike the effectual blow for economic freedom? How will we be freed from the industrial slavery which holds more than five million plow handle farmers and their families—twenty-five million men, women and children—bound to a bare existence, as their reward for bountiful crops so necessary for the very existence of the whole people?

How, brother farmers, will we free ourselves from economic conditions which continually drain the country of millions of dollars of hard earned cash and pile it up in the hands of a few greedy, selfish combinations of capitalists? Why was it possible for the beef trust to declare a dividend of thirty-five per cent; the oil trusts forty per cent; the sugar trust fifty per cent; the tobacco trust, sixty-two; the Adams Express Co., eighty per cent; the milk trust, one hundred twenty per cent, and so on up the column? How are we to account for this condition? By the simple fact that the people are not educated to be intelligent, moral, fraternal golden rule co-operators. The people are all powerful when educated and united.

The Farmers' Equity Union is an organizing, educating force. We are organized without capital or profit. Our exchanges have good financial standing in the business world of from ten to twenty thousand dollars, but the National Union has no capital. It binds together all other Equity Exchanges and leads to National co-operation.

We insist on putting a good weekly co-operative paper into the home of each member, which will educate him away from the old capitalistic idea and teach him and his entire family to be true co-operators. The Equity Union continually strives to promote the intelligence, morality and fraternalism of all of its members and to make them true co-operators. A continual effort must be made to overcome the ignorance, selfishness, narrowness and prejudice which now separates and weakens. The farmers can be educated. They are human beings with minds and hearts. They are educated on general subjects now, but must learn more about business and co-operative methods. They must be more fraternal. They must come together with a full determination to do right by each other and to know that their business is carried on honestly and efficiently.

Many co-operative enterprises in our country have failed. So many have failed that the pessimist says, "It can't be done." But most of them have failed for want of education, which leads to true co-operation. A strong majority in each community can and must be educated. The co-operative grangers of Johnson County, Kansas are a success. They organized their stock company in 1876 with a capital of $900 and now they are capitalized at one hundred thousand dollars. They have four large departments in their store in Olathe, Kansas, and a number of branches in Johnson County. They attribute their success for over thirty-five years to conservative management, wise, close supervision by the directors, but most of all to a continual campaign of organization and education among the thousands of farmers of that county. It is a practical demonstration of what education will do even among the farmers.

These co-operators say that the strength of co-operation is in the diffusion of its shares of stock, binding the interests of the whole community together and thereby holding their united trade. But its weakness is that this same is the legislator—the law maker of the concern, and though they all may be farmers— not one of them elevator men—yet the business must be managed and governed by the rules they adopt. Hence, the great importance of literature and school house lectures to educate a strong majority to follow the cooperative business plan of the Farmers' Equity Union.

Meetings and money are absolutely necessary. They are just as important to the Equity Union as to the church. Without these two essentials, churches and Equity Union would be a failure. One million farmers must be united under the Equity Union banner. They must meet once each month and be willing to put one dollar each year into the educational fund. This will furnish the sinews of war for battling down the great walls of prejudice, suspicion and selfishness which now make Union impossible. Every co-operative movement must have behind it this educating force. It is indispensible. We have two thousand so called elevator companies in our country. They must be educated away from the capitalistic system. It is only a question of time when all such elevators will be in the hands of a few capitalists again, unless we can get them on the cooperative plan. They must be induced to put in a by-law forbidding the Board of Directors to declare over five per cent dividends on the stock subscribed.

Big per cents on dollars is heathenish, and it is this idea or system that enables the selfish few to rob the toiling masses out of hundreds of millions of dollars each year. The co-operative idea is that the man is the unit and not the dollar. The capitalistic idea is that dollars are more important than men, women and children. Human greed squeezes out all love for humanity. Our only defense is in golden rule co-operation.

We want two hundred farmers at each good shipping point to get in exactly one hundred dollars each as their part of the working capital, build a good elevator and necessary warehouses and coalshed and run their own business. We are demonstrating at one hundred good towns that it can be done. This number will soon be doubled by the Equity Union. Every farmer who is able to put in twenty-five dollars can have the full benefit of co-operation. He can have all that he buys and sells at the elevator handled at the actual cost—without profit. The average farmer pays the profit-takers fifty dollars per year. Co-operation would put that fifty dollars back into his pocket instead of the pocket of Mr. Profit-taker, who perhaps is already a millionaire. Farmers, we have made millionaires enough by the profit system, it is time to change the system. Sacrifice, on the part of a few generous hearted, public spirited farmers will be necessary to start it at each good shipping point, to demonstrate that co-operation is right and will defend the wealth maker against the wealth taker. The concetration of a large volume of trade at each shipping point will reduce the cost of handling to the minumum. Co-operation will return to each patron's pocket all profits made on what he buys and sells. The Equity Exchanges will finally get all the business and be worth from ten to twenty thousand dollars every year at each point, and the farmers will wonder why they did not learn Equity Union co-operation before we made so many millionaires.

THE PRINCIPLES OF EQUITY UNION

The Farmers' Equity Union is organizing to carry into successful operation economic principles that are for the good of all humanity. It was started by three plow handle farmers of Illinois, but is destined to be the champion of every right and interest of the millions of American farmers as well as all other wealth producers of our Nation. Under the Equity Union banner must be gathered all classes of farmers who love justice and seek for themselves and all humanity equal rights and equal opportunity. If we adhere to just and righteousness principles from the beginning, "Public Sentiment," the "American King" will finally be for us and our success is assured.

Strong local Unions are being organized at the best shipping points in ten states, in which are united all classes of farmers and educators who are in sympathy with the movement. This should be a regular farmers' club, meeting regularly once each month. The educational feature must not be neglected nor over looked nor under estimated. We are organizing to promote intelligence, morality and fraternalism among our members, and to make them true co-operators. We are next to the churches and public schools in importance. We believe the prosperity and happiness of the people depend on golden rule co-operation. Ignorance, suspicion and selfishness separate and weaken the people and make them an easy prey for the greedy few who combine to rob them. Every link in our chain must be made stronger and stronger and all must be linked in the same great national chain. Wherever five or more farmers who believe strongly in this idea let them gather together, sign an application for a charter and send it to Greenville, Illinois, and begin to meet regularly and build up a local Union. Wherever ten or more will sign an application for a charter, elect a secretary and send three dollars for each member to the Farmers' Equity Union, Greenville, Ill., the National Union will send a lecturer and build up that local Union to one hundred members as soon as possible. We are especially desirous of organizing at good shipping points, where the farmers need a farmers' elevator. The one hundred or more successful Equity Exchanges are doing more to educate our farmers to be co-operators than all the writing, publishing and lecturing we can do, and yet the campaign of education must go on. We are organizing district Unions around Aberdeen, South Dakota, Oxford, Nebraska and Kansas City, for the purpose of doing a regular retail and wholesale creamery and mercantile business. The Equity Union is teaching the millions of farmers to do what the business world did long ago, viz., substitute co-operation for competition, both in buying and selling.

The whole movement is in its infancy, but has wonderful possibilities and promises rapid growth in the near future. After all the agitation in the old American society of Equity, we are just beginning to advocate true co-operative principles. Our organization is no longer a mere rope of sand, like the old Equity, but a powerful chain composed of strong links, and every link added on makes it more powerful. Any good shipping point in any state can form a link and share in the benefits of co-operation. Plans are being formulated to center all of each crop or product in one channel, controlled by the farmers instead of the grafters extortioners, profit-takers, speculators and combinations. Every member will read the same co-operative paper every week. This will lead to united action all along the line from Dakota to Texas, and from Ohio to Montana. Hundreds of the best organizers and lecturers on co-operation must be put into the field. Every member must be a local organizer.

One of our principle objects is to establish sure, steady, just prices for all farm crops and incourage and insure a full supply and reasonable prices to consumers. We do not believe in sixteen cent cotton in the spring when the speculators are selling it, and down to six cents in the fall when they are buying it from the farmers. We are opposed to $1.60 wheat in June and 80c wheat in July. To eleven dollar hogs one year and four dollar hogs an other year.

The Equity Union idea is for the steady demand to be met with a steady supply, and then we will have a steady price, which is just to both producer and consumer. When the farmers of the northwest own grain tanks on their farms, Equity Union elevators and terminals, they will hold their surplus grain in the hands of the producers, regulate the supply to the demand, and control the channel direct to the millers and exporters. No more low, unjust prices will be put on the poor producers, nor high extrotionate prices on the poor consumers. Mr. Speculator will be out of a job. No restraint of trade, no trust here, no co-operation of the few to the sorrow of the many. The principle is the greatest good to the greatest number. Equal rights to all.

When the stock men around Kansas City line up strongly in our local Unions, they will form the Kansas City stock men's district Union, own their own stock yards, and have their own commission firms. Then we will not see hogs eleven dollars one year and down to four in a few years, but a steady just price and full supply every month and every year. The beef trust will not be able to declare a dividend of thirty-five per cent, but hard working men in the cities and towns can eat meat three times a day and thus consume the steady increased supply, brought by the steady, sure prices. No violation of the anti-trust laws here—no restraint of trade. Production, consumption, sure prices, steady supply—all encouraged.

If the milk shippers around Kansas City will unite in our local Unions, with all other classes of farmers, we will establish a central market in Kansas City and organize the consumers with them and divide the profits on milk, cream, eggs, poultry, and produce between the producers and consumers. They will find behind them a great organizing, educating force which grows stronger every year and will insure success. When we as farmers unite our trade for machinery and take the entire output of individual factories, we will save millions of dollars each year. One million farmers united have more power than all the trusts combined. We must organize and control our patronage on what we buy and sell.

THE FARMERS' EQUITY A TRUST UNION

The Farmers' Equity Union is organizing a regular trust union, which will be approved by all fair minded people, when fully understood. We are uniting from one to three hundred farmers at each shipping point and educating them to trust themselves and one another in a co-operative business or Equity Exchange. At some places we start from a small beginning. Most of the great movements for the people have started in this way. We hold school house meetings and instill the true principles of co-operation. We make a house to house canvass by our organizers. We distribute literature and papers and educate our members to have confidence in themselves and to trust each other. We insist that each farmer must do right by his neighbor and be worthy of his confidence. We can do much by education.

Where one hundred farmers are organized into a local Union, they must be persuaded to each take one share of twenty-five dollars in the Equity Exchange. Then they can lease a ware house, put in a manager and start a co-operative business, handling eggs, poultry, calves, lambs and wool, and loading grain, hay, hogs and cattle in car load lots and also handling flour, feed, twine, salt, fencing, tile and fertilizer. This is a good way to start, but a canvass must be followed up from house to house and in school house meetings, until one hundred or more have each taken a twenty-five dollar share. Do not give it up till two hundred farmers are united and trusting each other in business.

Every thing must be bought and sold on a cash basis and at a fair profit. Stockholders may receive credit to the amount of the shares which they own. The Board of Directors must insist on a fair margin. The prices are never cut but must be the same as those of other dealers. Those handling money are sufficiently bonded and the books must be audited regularly every three months.

Every thing possible is done to make the business safe and successful, and to win and hold the confidence of the farmers. It is a trust union worthy of confidence. Running expenses, repairs, national dues and five per cent dividend come out of the gross earnings, and all net earnings are prorated to members according to patronage. No member can draw out anything until twenty-five dollars in full has accrued to his credit, when he is given another share and the capital increased by twenty-five dollars. As soon as he has four shares—the limit—cash will be prorated to him for his patronage.

As the farmers are organized and educated, and the capital and business are increased three hundred farmers will finally have exactly one hundred dollars each in the Exchange, or thirty thousand dollars invested and banked. All farm produce and wagons, and all farm machinery will be handled at actual cost. This big trust union will be on the principle of the greatest good to the greatest number, to which no fair man can object. Then, if they wish to increase the capital they can increase the limit of shares to five instead of four, and their patronage which they now throw away, will add $7,500 to their capital in one year. Do not have a sinking fund, nor a surplus fund. Increase your capital by increasing the limit of shares, and by your patronage, which you now give away to the profit system.

The patronage of three hundred good farmers for one year on all they buy and sell, at present profits, it worth $15,000 clear of the cost of handling. They must be organized and educated to conserve this patronage, which they now waste every year. They must agree to be brothers and trust each other. If thousands of them can be brothers in lodges and pay large sums of money for their support, they certainly can be induced to unite and be fraternal and support the Equity Trust Union, which will conserve their patronage to the extent of hundreds of millions of dollars each year.

The question which confronts every new movement is: "Will it pay?" The patronage of the farmers at one thousand good markets is worth more than fifteen million dollars annually under the present profit system. The new trust union would count that vast sum out in cash to the farmers at the one thousand points every year if they will organize on our plan, three hundred strong at every place. Brother farmers, will it not pay to lay aside our suspicions, fears, grudges, animosities, prejudices, and whatever separates us and unite in the Equity Trust Union? The business men co-operate in business whether they like one another or not.

If every wheat raiser was reading the Equity Union Exchange every week, every good bushel of the 1915 wheat crop would be sure to bring one dollar. We would trust each other and act together. All of our Exchanges would be under the daily direction of our own commission firm or expert salesman, which would be a decided advantage. All Exchanges would center their trade for flour and feed with the same mill, with the same mine for coal, with the same factory for twine, with another for wagons, another for automobiles and another for all farm machinery. Combination, concentration and co-operation will come as a result of brotherhood and fraternalism among our millions of farmers and will bring peace, plenty and prosperity and much real happiness.

The trust union is getting away from the capitalistic idea of declaring big dividends on dollars. We pay the man for his patronage instead of paying dividends on the millionaires' dollars. The man is the unit and not the dollar. When enough farmers are educated to trust themselves and to trust one another, we will knock out the profit system and break the power of every robber trust to harm the people.

When hundreds of Equity Exchanges are organized and co-operating together, then shipments of stock, hay, grain, etc., to all central markets can be so regulated each day by our commission firms that every central market will be fully supplied and none glutted, so that the steady demand for all food products will be met by a steady supply and prices of all farm products will be as sure and steady as on groceries, dry goods and farm machinery. This will mean hundreds of millions of dollars more for our best crops of grain and stock, and more reasonable prices to consumers. It will mean intelligent, distribution of our valuable products so that our largest crops of finest quality do not bring us less money than our small inferior crops.

It will prevent the drainage of hundreds of millions of hard earned cash, each year from the country into the hands of the greedy few. Brother farmers, have we not piled enough dollars into the great commercial centers by playing into the hands of speculators and trusts every time we have a good crop? Our distrust of each other is the cause of the whole business. We cannot afford it. It is time to lay aside competition for co-operation. To lay aside prejudice and distrust and come together as brothers in the Equity Union Trust. It will make us better church members and lodge members, better neighbors and citizens and better business men.

Let us organize a trust union and agree to no longer to trust speculators and profit takers to price our products, nor price-hammering middle men to run our business for us, but instead we will trust five good men as directors selected and controlled by us at each shipping point, to run our business for us and handle all we buy and sell at actual cost and thus pay thousands of dollars now sacrificed each year to uphold the profit takers. Such an arrangement, backed by the patronage of three hundred good farmers, is sure to be a grand success. Farmers, do not wrong your family, community and country any longer by refusing to trust each other and act together. Ten good farmers can start at your town. Will YOU be one of the ten?

THE EQUITY UNION PLAN SANE AND PRACTICAL

The Equity Union is advocating a sound, practical plan of co-operation. Its leaders are plow handle farmers, who have suffered heavy losses from failure of crops when prices soared high, and alike from low prices when crops were abundant and of fine quality. They are moved by honest motives and genuine sympathy for the condition of five millions of plow handle farmers, who are either renters or mortgaged home owners. The farmers starting the new Equity have had seven years' experience in their efforts to get the old Equity on a sane, simple, but comprehensive plan of co-operation. Our efforts failed in the old Equity but we are succeeding now in twelve states with the new Equity. In starting the new Equity, we have profited by our experience in the old, and we are cutting out all weaknesses and handicaps of the old American Society of Equity. We have a definite, practical plan of co-operation to offer every community of farmers. We know this plan will work and bring benefits to our members, as we are demonstrating its practicability at over one hundred good markets. We know definitely what we want to do and how to do it. Our object is true, practical co-operation by a very large per cent of the farmers at every good shipping point in buying and selling. When organized the business is run by five competent directors, who are under the control of the stockholders. The manager of each Exchange is sufficiently bonded and all books are audited regularly every three months by an expert bookkeeper. The Board of Directors are paid according to services and are expected to absolutely know how the business is running from week to week. The united patronage of three hundred or more farmers at our best shipping points on this plan insures the success of the business and is worth thousands of dollar annually to eac community.

The Equity Union believes in middle men, but we want them to be OUR middle men. We want them to be our well paid servants instead of our masters, each trying to make a fortune handling our business.

In one town which we visited, there were twenty-one men handling stock, hay, grain, feed, coal, flour, fertilizer, tile and farm machinery. A few of them have amassed a small fortune, and all of them have made a good living for their families out of the business. Their policy has generally been to hammer the price down on what they buy from the farmers. They buy as cheaply as possible and then combine sufficiently to hold up the price on what they sell the farmers, so as to support the whole bunch in town with its system of high-priced living. An Equity Exchange can use about five of the most honest, competent and industrious of the twenty-one in handling our business, and let the rest become producers or helpers instead of parasites. We will pay our middle men good living salaries for honest, competent service.

If we must support them, we want them to work for us and not against us. We want them to get the very best price for our products which the markets will justify. They must not be price-hammers, but price-boosters. Some of the most dangerous, dishonest and costly middle men are found in our central markets. Many of them are grafters and thieves. The most of them are there simply to execute orders and collect commissions. What we, as farmers, need is to realize that our principle trouble lies in the central markets.

The Board of Trade may not realize it, but the fortunes of this institution are being "weighed in the balance," and if it continues to allow itself to be used as a device for the reckless raiding of the price of farm products, the verdict of the country will be against it.

The Equity Union movement will give us real representation in the price making grain markets of the world. Our great need is a co-operative grain company with head quarters in Chicago and a branch in each central market. Every Farmers' Elevator Company should take one share in this company and have a voice in the control of its affairs. The object of this company should be to safeguard the interests of all Equity Exchanges in the interior markets through its agents and representatives, to supervise weights and inspection, to furnish information and advice covering all subjects pertaining to grain and grain products and to collect freight claims through its traffic department.

This company should direct each of our Exchanges daily, when and where to ship. A manager with such direction from a well informed company, could not go wrong. Speculation by this central company or any of its employees must be strictly forbidden in its by-laws and these by-laws enforced rigidly. Foreign connections should be established in Great Britain and on the Continent of Europe. Then shipments from the country to central markets will be directed and controlled so that each center will be fully supplied and no market glutted, and the present extreme fluctuations will be largely eliminated. This central co-operative grain company should be organized soon and handle the business of all Equity Exchanges.

This will give us expert salesmen to sell our great crops more directly to the consumers. A few brainy men who understand the markets of the world and have modern facilities for gathering information can easily distribute the wheat crop of the United States at a gain of more than one hundred million dollars annually, without raising the price to consumers. The expert handling of this crop would be an object lesson for other crops.

My brother farmer, let me speak to you privately: Are you one of the intelligent, well-to-do farmers, who are helping to defeat and hold back this grand movement for the American farmer by inaction, indifference, neglect or actual opposition? Why not right-about-face now and be the leader in your community for this progressive, up-to-date Union. We can make a strong link at your shipping point that will be worth thousands of dollars, annually to the farmers if you will co-operate with us in holding a few good rousing Equity Union meetings at your town.

THE EQUITY IDEA

The Equity idea does not belong to any one man, or to any set of men, and no one has a patent right or copy right on it. It is a growth among the people resulting from agitation and education, north and south, east and west, for seven years in the old Equity, in favor of co-operation.

The people are awakening to a sense of their danger from political bosses and millionair corporations and combinations, as well as to a sense of their rights and power in government and business when united.

The Equity idea is founded upon the principle of the greatest good to the greatest number. On equal rights to all and special privieges to none. This idea abhors captains of industry with their millions of industrial slaves.

It means economic freedom to all wealth producers, through organization, education and co-operation. It means rule by the people in politics and business and a substitution of a democratic for an absolutist industrial system. It will guarantee sure Equitable prices to the producer and more reasonable prices to the consumers. We have enlisted and won over to our side some of the leading agricultural papers of our country.

We have called the attention of commerce, state legislatures, and the government generally to this idea, and bills are being introduced against grain options, stock gambling and horse racing, and a commission has been appointed to probe the New York Stock Exchange. The new Equity Union idea has taken hold of the farmers in twelve states as no idea ever did before. They see in it economic freedom, self defense and protection against trusts and speculators who combine both in the country markets and leading centers and hold prices down on producers and up on consumers.

Newt Gresham started the farmers' educational co-operative union in Raines county, Texas in 1903, on this Equity idea. This organization has a large membership in the cotton states and low, unjust prices on cotton are beginning to be a thing of the past.

The new Equity idea is "Do not sell for a low price and do not hold for a high price." The promulgation of this doctrine for seven years in the best wheat state has resulted in the building of thousands of granaries on the farms and the holding the surplus in the country, and thus to some extent preventing low, unjust prices.

The new Equity Union is advocating the idea of complete ownership and control of the channel through which grain must be shipped to reach the mills, exporters, central markets and finally the consumers' kitchens. They are building their own elevators, warehouses, coal sheds, etc., in more than one hundred good markets. We are organizing the Equity Creamery and Mercantile Company at Aberdeen, South Dakota, Oxford, Nebraska and Kansas City, Kansas. Through the central markets we will finally reach the consumers direct with all farm produce, and eliminate all unnecessary middle men and save millions of dollars to both producers and consumers.

Every day shows progress in different states in the way of organizations looking to better markets and more intelligent systems of distribution. There is every indication that the farmers of our country are ready for a National Union that will unite all classes of farmers under one National Head, providing the organization is on a simple but practical plan. The plan must be simple—easily understood and easily carried out. Practical—showing benefits to every member, plainly, at his home market. Comprehensive—uniting under one National Head all classes of farmers in a National Union, and last, but not least, there must be sufficient revenue provided to carry on a continual campaign of organization and education, so that farmers may be led to see the power of organization and co-operation.

ADDRESS OF THE NATIONAL PRESIDENT DELIVERED AT THE STATE MEETING OF THE FARMERS' GRAIN DEALERS' ASSOCIATION, SPRINGFIELD, ILL., MARCH 7, 1911.

Ladies and gentlemen of the Farmers' Grain Dealers Association:

It gives me great pleasure to have this opportunity to address you on the subject of a National Union of farmers.

The necessity for this great organization is apparent to every intelligent, thinking farmer. The extreme fluctuation of the prices of farm products as well as the uncertainty of production makes the farmers' business ten times more hazardous than it ought to be. In the spring of 1904 cotton was 16c a pound and in the fall down to 6c a pound. In June, 1909, wheat was $1.60 a bushel and in July right after down to 80c a bushel. In September, 1907 and 1908 when we put up our hogs they were $7 per hundred and then down to $4.00 per hundred when the bulk of the crop went on the market in December and Jaunary. A prominent farmer from northern Illinois lecturered in Greenville at our Farmers' Institute. He said, "Feeding a bunch of cattle was like gambling on the Board of Trade in Chicago." Will some stock man rise up and tell me why it should not be just as sure and safe to put up one hundred steers and fatten them for the market, as it is to manufacture one hundred mowers or wagons and put them on the market.

The extreme difference between the farmers' prices and the consumers' prices would be ridiculous if it were not so serious. The poor producer and the poor consumer are at each end of the line, and the millionaire in the middle. Our costly system of distribution is partly responsible. In October, 1910, apples in Greenville were selling at $3.40 per bushel. I went southeast about 100 miles and found the farmers delivering good apples into the cars at 40c a bushel. I shipped a car load of those apples into Greenville and brought the price down from $3.40 per bushel to 75c a bushel. In January, 1910, I found the farmers at Coloma, Wis., were offered 20c a bushel for potatoes and they retailed in my town at 20c a peck. I found when near Appleton, Wis., that the farmers had sold a fine crop of cabbage for $5 per ton, and consumers in Indianapolis paid at the rate of $80 a ton. When we sell cheap hogs, wheat, milk, apples, beans, potatoes or cabbage as we often do, it does not mean cheap food to the people. When the poor producer in Nebraska sells cheap wheat, it does not mean cheap bread to the poor consumer in Chicago.

There is a steady demand for all farm crops which must be met with a steady supply; then we will have a steady price and a full supply. No shortage. The surplus must be held and controlled by the farmers instead of the speculators, then we will not see our largest, best crops sell for less money than the small, inferior crops.

A National Union of farmers can reduce the cost of distribution of crops thirty per cent, and the cost of farm machinery fifty per cent on the co-operative plan of the Farmers' Equity Union. This Union is chartered by the Secretary of State, Springfield Illinois, with headquarters at Greenville, Illinois. The charter is dated December 16, 1910. It is organized without profit and without capital. It is an organizing educating force—a force which must be behind every Farmers' Elevator Company to teach true co-operation and insure success. A weekly co-operative paper is put into every member's home for four years when he joins. This paper teaches copoerative buying and selling. Lecturers and organizers are kept busy holding school house meetings every night in the week and at the town on Saturday afternoon. This Union continually promotes the intelligence, morality and fraternalism of its members and has for its chief object co-operative buying and selling to the advantage of all of its members. It treats outsiders fairly and holds its doors wide open for all farmers of good character to enter and enjoy its benefits and protection.

Local Unions or farmers' clubs are organized at every good shipping point and meet at least once each month. Then these Unions are organized into Equity Exchanges. The shares are $25 each; and the limit four shares to each member. Only members can be stockholders. Our Exchanges handle grain, hay, stock and all farm produce, also flour, feed, salt, twine, tile and all farm machinery. We try hard to run on a safe margin the same as the markets around us, so that there will be no assessments necessary. The fundamental principle of Equity is, that not over five per cent dividends can be declared on the stock subscribed. The By-laws forbid it in every stock company. Out of the gross earnings we take the running expenses and necessary repairs. The balance is net earnings and is prorated back to stockholders according to patronage. We will unite the trade of Greenville, Ill. farmers until we center a trade of three hundred thousand dollars worth of business each year at that Exchange.

With the present margin we can easily prorate five per cent of this amount, or fifteen thousand dollars each year to our three hundred stock holders when our plant is properly equipped. Then each stockholder who has furnished one thousand dollars worth of the business will receive fifty dollars, as his patronage dividend. The one hundred dollars he owns in the Exchange is simply his part of the working capital. The fifty dollars which he receives is his patronage dividend and will keep him in the Union from year to year. Every stockholder who has patronized us to the extent of five hundred dollars would receive twenty-five dollars for his patronage, in cash if he has four shares, the limit. If he has less than four shares we give him another share and add the twenty-five dollars to the capital. We will have three hundred farmers at Greenville, Ill., united, each having exactly one hundred dollars in the Exchange, with a capital of twenty thousand dollars invested in the business. We will pay non-members as much as members, and sell to them as cheaply but we will not prorate to them anything until they become members and stockholders.

We can lick every trust in the country if we get enough farmers united. But we will first have to lick some farmers into line. If a farmer finds that year after year his Equity neighbors receive back from ten to fifty dollars for their patronage, he will finally come in, become a member and take stock. The plan bids for members, stockholders and patronage. A very large per cent of the farmers at each good shipping point will be united on this plan of golden rule co-operation.

Strong local Unions at every good shipping point with an Equity

Exchange which handles the members' business co-operatively, without profit is the foundation of a grand national movement that will unite millions of farmers and bring economic freedom to the people.

THE EDUCATIONAL FEATURE MUST NOT BE OVERLOOKED, AND THE BUSINESS ORGANIZATION IS EQUALLY AS IMPORTANT. THE TWO MUST GO HAND IN HAND.

There must be an organizing educating head—in other words, a National Head—which carries on a continual campaign of organization and education without which there can be no lasting, permanent success. Tons of literature must be printed and scattered. The Equity Union Exchange must reach the home of every member weekly; this paper must teach not only scientific production, but co-operative buying and selling as well. Both are important, but the latter much more than the former. Organizers must be put into the field in increased numbers, as our means increase by the development of the Union. The lecturers, literature and the co-operative paper must concentrate their efforts in establishing Exchanges, through which the members can see practical demonstration of true blue co-operation. Never allow over five per cent dividends on the stock subscribed. Prorate to the members all net earnings according to patronage. Figure the net earnings as a per cent of the business furnished by the stock holders. If you have three per cent net earnings, it belongs to the patrons who are stockholders and should be prorated back to them according to the patronage furnished by each stockholder. Only patrons who are stockholders will be able to reap any benefits from our co-operation. No danger of the capitalist sneaking in to buy up the stock. He is out of a job. The people are their own capitalists. They have learned golden rule co-operation and do not need Mr. Millionaire Capitalist. Poor old moneybags! He will be compelled to work for his living. "He that will not work must not eat," is a bible truth, and applies to the idle rich as well as to the idle hobos.

The possibilities in this movement are beyond comprehension. It means our own terminal elevators, central stock yards and wholesale distributing plants in all of the large cities for the distribution of milk, eggs, poultry, fruit and all produce. It means that the farmers will control the channel direct to the consumers instead of the speculators, grafters, profit takers, extortioners and millionaire trusts. It means the cost of distribution of our crops reduced from thirty to fifty per cent; it means sure, steady prices to producers and reasonable prices to consumers. True co-operation will bring economic freedom to the American people. The two thousand farmers' elevators must be united under one National Head and be educated to become true blue co-operators. Three thousand new Exchanges must be organized as rapidly as possible. This plan will unite more than two hundred farmers at each good shipping point, and we will have one million farmers united and all true blue co-operators. What power and protection this would give every toiling yoeman in this grand country. Millions of dollars now being centered in the hands of millionaires would come out into the country to improve the soil, beautify country homes, schools and churches, build good roads and pay for automobiles. Country life commissions appointed by the President would be unnecessary; boys and girls would gladly remain on the farm.

The organization of our patronage will reduce the cost of farm machinery fifty per cent. When one thousand links are made and linked together in the Equity Union chain we can go to a wagon factory and buy its entire output for one year at actual cost of labor, material and transportation. We can then distribute wagons to our members at fifty dollars each which now costs from seventy to ninety dollars. We can furnish at retail the very best eight foot self-binder for $75 instead of $150, and we can buy the very best automobiles for one-half the present price. We will be equipped to distribute farm machinery with our one thousand ware houses at the best country markets. We will organize into a strong stock company at each good country market with financial standing in the business world. We will go to the factory with our patronage organized, then there will be no necessity for spending thousands of dollars for advertising and traveling men and no risk of sale. Our factory will run full time with full assurance of the sale of every wagon, machine or automobile. We take that factory out of the competitive world, with all its expense, risk and worry, over into our co-operative world of safety, peace and plenty. We will not go into the manufacturing business, but organize and center our patronage with individual factories. We will not need Sears & Roebuck nor Montgomery Ward & Co. to stand between us and the factory, as now at a profit of six or eight million dollars each year.

NATIONAL CONSTITUTION AND BY-LAWS OF THE FARMERS' EQUITY UNION.

Article 1. Section 1: Name and membership—This organization shall be known as the Farmers' Equity Union. It shall consist of farmers, editors, teachers, preachers and other educators who favor the accomplishment of the purpose of this Union, and who shall be accepted therein according to the prescribed rules of receiving members.

Sec. 2: Objects—The objects of this Union are to promote intelligence, morality, sociability and fraternalism among its members and to secure fair dealing in all the business relations of farm and mercantile life, and its purposes are fully set forth in the articles of incorporation, the chief of which is co-operation in buying and selling all products of the farm and machinery, automobiles, groceries, dry goods, clothing, and every household necessity. Co-operation to the advantage of all of our members is our chief object. We work for benefits for our members.

Article 2. Section 1: Organized Forms.—The organized forms are local and district Unions and a National Union, which is the Supreme Head of this organization.

Article 3. Section 1: Local Union.—Local Unions shall consist of farmers and others here before specified, accepted into the Union according to its rules and usages.

Sec. 2: Exclusive Control.—Each local Union shall have exclusive control of its own business affairs, and may adopt by-laws not in conflict with those of the National Union.

Sec. 3: How To Organize.—To organize a local Union, at least five persons qualified for full membership may assemble of their own accord and proceed to organize themselves into a local Union, by each paying an entrance fee of $2, and one dollar subscription for the paper, signing an application for a charter, electing the officers required and making due report to the national secretary with the remittance of three dollars for each member. Or if possible, a commissioned organizer should be called to organize.

Sec. 4: Entrance Fees.—Every person joining this Union as a regular member shall pay an entrance fee of two dollars, and one dollar for the paper. This three dollars shall be sent to the National Union secretary by the local Union secretary and shall be used for the promotion, spreading and building up of the organization. Every member joining must become a subscriber to the Equity Union Exchange, the official paper.

Sec. 5: Dues.—Every regular member shall pay to the National Union one dollar a year dues, payable in advance November first of each year. Local Unions shall fix their own dues.

Sec. 6: Special Members.—Special members are the wives and minor sons and daughters of regular members. They are admitted free but may pay local Union dues of five cents per month.

Sec. 7: How Charters Are Granted.—On receipt of an application for charter by properly organized local Unions accompanied by the entrance fee of two dollars for each member and one dollar for the paper, the national secretary shall make proper record thereof under the name chosen by the charter members and the next consecutive local union number, and shall transmit to the secretary thereof a charter duly and properly executed and attested by the seal of the National Union.

Sec. 8: Demits.—Any member in good standing wishing to change his or her membership may by paying up all arrearages and by a majority vote of his or her local union, be granted a demit for that purpose, of which transfer the local union secretary must notify the national secretary at once.

Sec. 9: Officers.—The officers of a local union shall be a president, vice-president, secretary-treasurer and a lecturer. The official term shall be twelve months and the annual election shall be by ballot in December of each year. All officers shall serve until their successors are elected and qualified.

Sec. 10: Time of Meeting.—The first Saturday of every month is Farmers' Equity Union day, and every member is under obligation to quit work and to take his family to the meeting at 2:00 p. m. in October, November, December, January, February, March and April, and at 7:30 p. m. in May, June, July, August and September. Provided local unions may change the time of their meetings.

Sec. 11: Committee on Programs.—The president shall appoint a committee on pragram of music, declamations, readings, recitations, debates, papers and speeches for each meeting. The local union shall be a regular farmers' club promoting the intelligence, morality, and every social interest of the farmers and their families.

Sec. 12: Duties of Officers.—The president shall preside at all meetings, shall maintain decorum, and see that laws of the Union are enforced. He shall fill all official vacancies by appointment pro tem in each meeting.

The vice-president shall assist the president in his duties when called upon and in president's absence he shall perform the duties of that station. In the absence of both president and vice-president, the secretary shall preside at all local meeting.

The secretary-treasurer shall keep a correct record of all proceedings of the local union, including in the minutes of each meeting, a statement of all payments of money by the members at that meeting. He shall keep correct list of all the members showing date of joining, who are regular and who are special members. He shall keep a day book account with the local union showing receipts and expenditures, and make a report of the same in open meeting once each month, showing balance in the treasury. He shall also keep a ledger account with each member giving him credit for each payment made to the Union. He shall collect all the entrance fees and forward the same to the national secretary with the subscriptions for the official paper. It shall be his duty to collect regularly and promptly all dues from both regular and special members. He shall deposit all money received in a bank and pay out no money except by bank draft or check. Before entering upon his duties he shall give a good and sufficient bond. The compensation of the secretary-treasurer shall be fixed by each local union.

The lecturer shall be the chairman of the committee on program each month. He shall see to it that teachers, professors, editors, lecturers and educators are invited to address the union meetings. Corn shows debates, lectures and declamation contests are recommended to keep up intrest.

Sec. 13: A Quorum.—Five regular members must be present to constitute a quorum provided every member has been notified of the meeting.

Section 14: Special Meetings.—The president or secretary may call a special meeting of the local union. The president shall call a special meeting at the request of ten or more regular members, but all regular members must be notified of the meeting, its time, place and object or objects. Only business mentioned in the call can be transacted.

Sec. 15: Not a Lodge.—This is not a lodge or secret society with ritual, pass words and grips, and shall not be in the future. But all the business councils and transactions shall be private and kept by the members as a protection to the business interest of the Union.

Article 4. Section 1: Offenses.—Local Unions shall have power to deal with their members for offenses against the union, and shall be governed by the rules usually applied in such cases. In case of conviction to be determined by a vote of guilty or not guilty, punishment may be reprimand, suspension or expulsion, also to be determined by a majority vote, voting first upon the severest penalty, if that fails to carry, then on the next and so on until the penalty is fixed. If no penalty is fixed, the president shall dismiss the case. Suspension shall not exceed three months and the national secretary must be notified of expulsion. Appeal may be taken by either to the National Board of Directors and their decision shall be final.

Sec. 2: Officers neglecting to attend meetings twice in succession will be sufficient cause for the Union to vote on declaring the office vacant, the majority to decide. No member shall have a right to vote in any meeting who is in arrears for dues or fines. Any member in arrears may be reinstated by paying up in full, provided he is not more than six months behind, in which case he must come in as a new member.

Sec. 3: Equity Exchanges.—As soon as practical each local union shall organize an Equity Exchange. Only members of the Farmers' Equity Union shall be allowed to take stock. The shares shall be $25 each and the limit four shares. The Exchange must be chartered by the state in which it is located. All farm produce including live stock may be shipped out. Coal, flour, feed, salt, cement, fertilizer, twine, fencing, groceries, wagons, machinery, automobiles, etc., may be shipped in. All shall be bought and sold on a safe margin. A board of five directors shall have charge of the business and shall hire a good manager. Out of the gross earnings shall be paid the running expenses, one dollar per annum for each regular member's national dues, provided he is a stockholder, and not over five per cent dividend shall be declared on the stock subscribed. The running expenses and national dues must come out of the gross earnings and all over this shall be net earnings and shall be prorated among the stockholders, according to the amount of patronage given during the year. The net earnings shall be figured as a per cent, of the business furnished by the stockholders during the year. Five thousand dollars net earnings made on one hundred thousand dollars of business furnished by the stockholders gives a patronage dividend of five per cent and each stockholder whose patronage amounted to five hundred dollars during the year would receive twenty-five dollars in cash for his patronage. Those whose patronage amounted to one thousand dollars would receive fifty dollars, etc., provided nothing shall be paid back until he has four shares, the limit. Pay him shares instead and increase the capital of the Exchange. Pay non-members as much for their produce as members and sell to them as cheaply, but give them no part of the net earnings until they become members and stockholders. The Exchange will handle every members' produce and merchandise at actual cost, giving back all profit. It will make a difference between members and non-members. On this plan you bid for members, stockholders and for patronage. You bring a large volumn of trade together to one center which reduces expenses and insures success. You knock out the profit system, which is a robber system, and introduce the co-operative system, which is the economic salvation of the farmers and all wealth producers. The directors shall not declare over five per cent on the stock subscribed.

Article 6. Sec. 1: National Union.—As the National Union is the Supreme Head, the parent organization, and every local is a child, the parent organization must receive liberal financial support from every local Union, with which to carry on a continual campaign of organization and education so absolutely necessary for the growth and life and success of this grand movement among seven million farmers and their families.

Sec. 2: Representation in National Union.—The National Union is the great combining, organizing force and shall consist of its officers, members, standing committees and representatives from the local union or Equity Exchanges. Every local union or Equity Exchange shall be represented at the National convention by its president, vice-president or secretary-treasurer of either the union or of the Board of Directors of the Equity Exchange, and by the manager of each Exchange, these representatives being expected to represent every material, agricultural interest of the country, including grain, live stock, dairy products, wool, cotton, fruit, vegetables, poultry, etc. poultry, etc.

Sec. 3: Meetings.—The National Union shall meet annually in December at a precise time and place fixed by a National Board of Directors. Special meetings may be called by the National President, or the National Board of Directors. Only subjects embraced in the call for special meetings shall be considered at such meetings and the National Secretary shall notify every local secretary, sixty days or more before said special meeting.

Sec. 4: Officers.—The officers of the National Union shall be a president, vice-president, secretary, treasurer and a Board of Directors. The Board of Directors shall consist of one director from each state in which this Union is organized, each director to serve four years. The secretary and treasurer shall be appointed or employed by the National President subject to the National Board of Directors. Each year at the annual meeting the National Convention shall elect a president and a vice-president from the Board of Directors.

Sec. 5: Duties of Officers.—The National President shall give his whole time and very best efforts to spreading and building up the Union. He shall preside at all meetings of the Board of Dirctors and of the National Union. He shall have supervision of the work of the Union in the absence of the National Board of Directors. He shall inaugurate, superintend and carry on a continual campaign of organization and education with a view to organizing and building up local Unions and Equity Exchanges. He shall also have full charge of the National Headquarters.

The vice-president shall be a regular member of the National Board of Directors, and shall perform the duties of the president in his absense or incapacity for any cause. The duties, bond and compensation of the national secretary shall be prescribed and fixed by the National President, subject to the orders of the National Board of Directors.

Article 7. Sec. 1: The National President shall engage lecturers and organizers together with all assistants required to properly conduct the work of the National Union and of the National Headquarters.

Sec. 2: The Board of Directors shall meet from time to time as they deem necessary. The National President may call meetings of the Board of Directors, or upon the written request of three members of the board, he must call a meeting. Three shall constitute a quorum of the Board of Directors, provided every member has been notified of the time, place and object of the meeting.

Sec. 3: The Board of Directors shall be a standing committee, to revise and recommend changes in the constitution and By-laws, provided they shall receive and consider any change or changes recommended or suggested in writing by a local Union, and they shall only have power to recommend changes to the National Union for its action.

Sec. 4: Changes in the constitution and by-laws may be made by a majority vote of the National Union in the annual meeting or at special meetings called for that purpose.

Sec. 5: Official Salaries.—Members of the Board of Directors shall be compensated only for such time as they are actuallly in the service of the Union at the rate of $3 per day and necessary traveling expenses. The National President shall receive one thousand dollars per annum and all expenses necessary for traveling, advertising and organizing purposes.

Sec. 6: The Equity Uuion Exchange shall be the official paper for the Farmers' Equity Union, and every member joining the Union, must be a four year's subscriber to said official paper at twenty five cents per year.

BY-LAWS

Article 1. Sec. 1. General Provisions.—Every effort must be made to organize and build up strongly local Unions and Equity Exchanges for co-operative buying and selling. The official paper, the textbook, the organizers, lecturers, national and local Union officers and all members must combine their efforts in this one direction. Every member is expected to be an organizer and an educator.

Sec. 2: Women owning farms may become regular members, provided any woman may become a regular member, who pays the regular entrance fee of three dollars.

Sec. 3: Every person handling Equity Union money must be required to give a good and sufficient bond and to make a monthly financial statement. All money must be banked before being payed out.

Sec. 4: Any woman who is a regular or special member may be eligible for secretary and treasurer of a local Union or of the National Union.

Article 2. Sec. 1: The discussion of partisan or secretarian questions is forbidden in all of our meetings and members may vote in politics as they choose.

Sec. 2: Amendments.—This constitution may be amended by a majority vote at any regular meeting of the National Union or at a special meeting called for that purpose.

BY-LAWS OF ....

EQUITY EXCHANGE

Article 1. Sec. 1: Name.—The name of this corporation shall be the .... Equity Exchange.

Sec. 2: Object.—This Exchange is organized to buy and sell all products of the farm, also farm machinery and merchandise of all kinds.

Sec. 3.—Capital Stock. The capital stock of this corporation shall be .... dollars, divided into .... shares of twenty-five dollars each.

Sec. 4.—Seal. The corporate seal shall contain the full name.... .... Equity Exchange.

Article 2. Sec. 1.—Directors. The business of this corporation shall be conducted by a board of five directors elected for one year who shall serve till their successors are elected.

Sec. 2.—Managers and stockholders in the said Equity Exchange must be members of the Farmers' Equity Union.

Sec. 3. The stockholders shall elect the president, vice-president and secretary-treasurer of this Equity Exchange at their annual meeting the last Saturday of June of each year, and these officers shall be the local union officers. These officers shall serve until their successors are elected and qualified.

Sec. 4. Vacancies in the Board of Directors shall be filled by the stockholders at special meetings called for that purpose or at the annual meeting.

Sec. 5. The directors are authorized to employ a manager, and all necessary help to carry on the business successfully. They shall fix the compensation of all officers and employees, provided the members of the board shall only be paid for actual service at the rate of forty cents per hour.

Article 3. Sec. 1.—Duties of Officers. The president shall preside at the meetings of the directors or of the stockholders. He shall sign all certificates of stock, call special meetings of the directors or stockholders when he deems it necessary or when twenty per cent of the stockholders petition for a meeting. He shall sign all bonds, contracts or other instruments in behalf of this Exchange when so ordered by the directors.

Sec. 2.—Vice-president. In case of absense of the president, or when called upon to serve, the vice-president shall perform the duties of the president. He shall be a member of the Board of Directors.

Sec. 3.—Bonds. All officers and employees handling the money of this Exchange shall be sufficiently bonded and the bookkeeper's books shall be audited regularly as often as the directors deem necessary.

Sec. 4. The secretary-treasurer shall keep correct minutes of all meetings of the directors and stockholders, have charge of seal, records, books and assets of the corporation, subject to the orders of the directors. He shall sign all certificates of stock and attach the seal there unto.

Article 4. Section 1.—Not Over Five Per Cent. Dividends on Stock. The Board of Directors are authorized to pay the running expenses and all necessary repairs out of the gross earnings of the company, and to use the capital to make necessary improvements. They are prohibited from declaring over five per cent dividends on the stock subscribed. Out of the gross earnings they shall take running expenses and necessary repairs and also the national dues of each stockholder in the Farmers' Equity Union and not exceeding five per cent dividend on the stock subscribed. All earnings over this shall be net earnings and shall be prorated back to the stockholders according to patronage. The net earnings or profits shall be figured as a per cent of the business furnished by the stockholders during each year.

Sec. 2. No money shall be drawn from this company by any stockholder until he has four shares, the limit. He shall be given shares instead and the capital increased by the amount.

Sec. 3.—Fraud. No stockholder shall market any other farmers' produce as his own nor attempt to give any outsider the benefits of co-operation. For each offense he shall be fined one hundred dollars. Provided, if the tenant is a stockholder and the landlord not, the tenant may market the entire crop as his own and receive prorations on the entire crop. The landlord must not be given any benefits of co-operation until he becomes a stockholder. Non-members of the Union must not be given any benefits of co-operation because they cripple our cause.

Article 5. Sec. 1.—All elections shall be by ballot.

Sec. 2.—Quorum. A majority of the directors shall constitute a quorum, and twenty-five per cent of the stockholders shall constitute a quorum at their meetings.

Sec. 3. All orders, minutes of meetings and stock certificates shall be signed by both president and secretary.

Article 6. Section 1.—Safe Margin. The Board of Directors shall insist on a safe margin in buying and selling, and prorate all net earnings to stockholders according to patronage, paying cash to each stockholder who has the limit of shares, and paying in shares all who have less than the limit.

Sec. 2. The directors shall carry on a continual campaign for more members of the Union and stockholders and thus increase patronage and capital.

Sec. 3. By a majority vote of the stockholders at a regular meeting the limit of shares may be increased when more capital is needed.

Sec. 4. The regular annual meeting of the stockholders shall be the last Saturday in June of each year, when the Board of Directors shall make a full report of the business of the company for the past year, and prorate all the net earnings back to stockholders according to patronage.

Sec. 5. The directors and managers shall make a cut off in January and June of each year and all business of the Exchange and books must be audited for six months by an expert auditor.

Sec. 6.—Complaints. All complaints by stockholders shall be made to the directors in writing, and signed by the complainant. The directors shall make such investigation and decision thereon as they shall deem proper, subject to an appeal to the next regular meeting of the Exchange, which decision shall be final.

Sec. 7. In November of each year the Board of Directors shall pay the national dues of each stockholder and charge the same to his account.

Sec. 8. Each stockholder shall have one vote and only one, and no one shall hold over four shares in the Exchange, unless the limit of shares is changed.

Sec. 9. This Exchange shall send its president and manager as delegates to the annual meeting of the Farmers' Equity Union each year and pay their traveling expenses.

Sec. 10. These by-laws may be amended by a majority vote at any regular annual meeting of stockholders, or at a special meeting called for that purpose.

THE EQUITY CREAMERY & MERCANTILE EXCHANGE BY-LAWS

Article 1. Sec. 1. The name of this corporation shall be, "The Equity Creamery and Mercantile Exchange."

Sec. 2. The place of business shall be in........................

Sec. 3. This Exchange is organized to do a general wholesale and retail creamery and mercantile business.

Sec. 4. The capital stock of this Exchange shall be fifty thousand dollars, divided into five thousand shares of ten dollars each.

Sec. 5.—Seal. The corporate seal shall contain the full name, "The Equity Creamery & Mercantile Exchange."

Article 2. Sec. 1. Only members of the Farmers' Equity Union can take stock in this Exchange.

Sec. 2. This Exchange shall be made up of stock holders, and a board of five directors.

Sec. 3. There shall be an annual meeting of the stockholders of this Exchange in............................., on the last Tuesday of November at which meeting all stock may be represented by the owner or by proxy.

Every local Union of the Farmers' Equity Union, having three or more members who are stock holders in this Exchange may send one delegate to the annual stockholders meeting at the expense of the Equity Creamery & Mercantile Exchange, who may vote any proxies held by him of the members of his local union.

Provided any stockholder may attend the annual meeting of the stockholders at his own expense and cast his own vote. No unsold stock shall be voted. These provisions shall apply to special meetings of the stockholders.

Sec. 4.—Officers. The business of this Exchange shall be conducted by a board of five directors, elected by the stockholders annually.

These directors must serve till their successors are elected and qualified.

Sec. 5. One director shall be elected president and one vice-president in each annual meeting and shall serve until their successors are elected and qualified. The directors shall elect one of their number secretary-treasurer.

Sec. 6. In the first meeting the directors and officers shall be chosen temporarily until the last Tuesday in November, 1914.

Sec. 7. The last Tuesday in November of each year shall be the regular annual meeting day of the said Equity Creamery & Mercantile Exchange.

Sec. 8. At the first regular annual meeting the regular delegates and other stockholders present, shall organize permanently by electing a board of five directors for one year and electing one of said directors president and one vice-president.

The directors shall elect one of their number secretary-treasurer.

These officers shall serve until their successors are elected and qualified.

Sec. 9. Vacancies in the Board of Directors shall be filled by the directors.

Sec. 10. All meetings of directors or stockholders shall be held in .... at a place fixed by the directors.

Sec. 11. It shall be the duty of the president of this exchange to give his whole time and very best efforts to organizing and building up this Exchange in the territory tributary to ....
......................He shall superintend the entire business of the Exchange, subject to the orders of the directors.

For his services he shall be paid one thousand dollars per annum salary, and all necessary expenses in traveling, advertising meetings and organizing this Exchange.

Sec. 12. The vice-president shall serve in the absence of the president or when called upon.

The secretary-treasurer shall keep correct minutes of all meetings of directors and stockholders, have charge of seal, records, books and assets of this Exchange, subject to the orders of the directors. The directors shall fix his compensation.

Article 3. Sec. 1. All certificates of stock must have the seal attached and be signed by the president and secretary.

Sec. 2. The directors are authorized to employ a butter maker, manager, bookkeeper and all necessary help to carry on the business successfully. They shall fix the compensation of all officers and employees except the president, provided the directors shall be paid three dollars per day and necessary expenses for actual service.

Sec. 3. All regular delegates to regular or special meetings of this Exchange shall have their rail road fair and hotel bills paid by the said Exchange, providing they attend the meetings.

Article 4. Section 1.—Distribution of Earnings. The Board of Directors shall insist on a safe margin in buying and selling and prorate all net earnings to stockholders according to patronage, paying cash to each stockholder who has two shares, the limit, and paying in shares all who have less than the limit.

Sec. 2. The Board of Directors are authorized to pay the president's salary and expenses as provided in these by-laws, and also the running expenses and necessary repairs out of the gross earnings of the Exchange, and to use the capital to erect or buy buildings and equipment and to make improvements.

They are prohibited from declaring over three per cent stock dividends. All earnings over this shall be net earnings and shall be prorated back to stockholders according to patronage. The net earnings shall be figured as a per cent of the business furnished by the stockholders each year and paid back as a patronage dividend.

No money shall be drawn from this Exchange for patronage by

any stockholder, until he has two shares, the limit. He shall be given shares instead, until he has two shares.

Sec. 4. No stockholder shall market any other person's produce as his own nor attempt to give any non-member the benefits of co-operation. For each offense he shall be fined one hundred dollars.

Article 5. Sec. 1. All elections shall be by ballot.

Sec. 2.—Quorum. A majority of the directors shall constitute a quorum and twenty-five regularly appointed delegates may meet and transact business legally, provided all stockholders have been notified by three insertions in the Equity Union Exchange, our official paper.

Sec. 3 All orders, minutes of meetings and stock certificates shall be signed by both president and secretary.

Sec. 4 The president of this Exchange shall carry on a continual campaign for more stockholders and thus increase patronage and capital.

Sec. 5 By a majority vote of the representatives at a regular meeting, the limit of shares may be increased if more capital is needed. Provided all stockholders are notified through the paper, "The Equity Union Exchange," by two insertions that, that question will be voted upon.

Sec. 6. At the regular annual meeting on the last Tuesday of November of each year, the Board of Directors shall make a full report of the business of the Exchange for the past year, and prorate all net earnings back to the stockholders according to patronage.

Sec. 7. This Exchange shall send as delegates to the National meeting of the Farmers' Equity Union its president and secretary, each year and pay their necessary traveling expenses.

Sec. 8. These by-laws may be amended by a majority vote at any regular annual meeting of stockholders or at a special meeting called for that purpose.

A NATIONAL UNION

The great need of a National Union of farmers is more apparent every day. The combinations which rob the people grows stronger continually. They pool their interests, merge their banks and combine their forces more every year. Their is more Union and co-operation of the few to the sorrow of the many each decade. The billions of dollars of wealth produced annually by the millions of farmers and their families each year go to strengthen the combined forces of the money kings. The farmers are producing more real wealth than any body else and piling most of it into the hands of the trusts.

The grain growers enrich the speculators and milling trusts, the dairymen the milk trust, the potato men the potato kings, and all classes of farmers are robbed by the machine, lumber, sugar, coffee, salt, shoe and clothing trusts. We are robbed on what we sell and what we buy. The entire system through which we buy and sell is a robber system, supporting thousands of unnecessary middle men. The profit system in vogue today is the curse of the farming fraternity. Big profits on what we buy and sell make the millionaires.

The distribution of our wealth to the people who produce it is the chief object of the Farmers' Equity Union. This is the great problem before the American people, and while political parties may assist, it will have to be solved chiefly by industrial Unions, which teach the people to be more intelligent, moral and fraternal, and true blue golden rule co-operators.

The Farmers Equity Union will unite under one National Head cotton, grain, stock, dairy, fruit and produce men, all co-operating for their mutual interests.

The foundation of this Union is strong local Unions at our best shipping points. Every farmer should be induced to join a local Union, and the fact that he is a regular member should make his entire family special members.

The local Union should be a regular farmers' club, held the first Saturday of every month in day time in winter and at night in the summer. The first Saturday of each month is Equity Union day all over the United States, and must be religiously observed by every member and his family. We must not be too busy in the field to go to our meetings.

The half day spent in meetings each month will be worth more to us than all the other days spent in the fields. Every regular member must pay one dollar November first each year for the support of the National Union and local Unions must fix their own dues. The organization which attempts to run without money is dommed to failure. The national dues must be paid out of the proceeds of the Exchanges and must be used economically to support a good national headquarters, a good co-operative paper both in English and in German, and especially to keep in the field the very best lecturers obtainable. A great campaign of organization and education is an imperative necessity. This is impossible without money. Thousands of dollars must be spent for education, and the result will be millions of dollars in the farmers' pockets, instead of in the millionaires' unholy coffers.

A good co-operative paper must reach every member's home weekly. Every local Union must have a correspondent for this paper, giving all encouraging news, increase of membership, results accomplished by co-operation and discussions and suggestions for the improvement for our great Union. Let us fill a sixteen page paper full to overflowing every week with discussions of the great economic questions now before the American people and especially with all of the encouraging news favoring co-operation which we can send in.

The local Union president should appoint a committee on program, who should see to it, that music, declamations, readings, debates and speeches are provided for at each meeting.

A regular lecture course will finally be provided for our Unions each year, and our finest quartets of singers will tour the country, singing Equity Union songs, which will arouse and educate farmers by the thousands. We must sing the gospel of Equity Union as well as to preach it.

The organizing, educating feature must not be neglected nor overlooked in this grand movement, if victory is to be ours. Farmers can be educated to become golden rule co-operators in business, and this is the goal we wish to reach, viz: Economic freedom in the business world. When one thousand of our Exchanges have financial standing in the business world and are co-operating, a new system of distributing our products will be developed and wholesale buying agencies established worth millions of dollars to our members.

Every farmer who reads this should enlist at once, not for the summer, "but until the end of the war," for manhood and patriotic service will be needed in this conflict.

I quote here the prophetic words of an ex-governor of Illinois, a leading democratic: "But, some one says, is there any use in our making an effort?" Are not all the bankers of this country, all of the trusts and great corporations of this country, all of the powerful forces of this country, is not the fashion of this country are not the drawing rooms and the clubs of this country now controlled by concentrated and corrupt wealth? Are they not growing stronger every year, and do they not villify and attempt to crush every body that does not submit? Can anything be accomplished in the way of curbing this great force and protecting the American?

"My friends, let me site you to a parable: George William Curtis and other writers of his day have described the slave power back in the fifties. They tell us that slavery sat in the white house and made laws in the capitol; that courts of justice were its ministers. That senators and legislators were its lackeys; that it controlled the professor in his lecture room, the editor in his sanctum, the preacher in his pulpit; that it swaggered in the drawing room; that it ruled at the clubs; that it dominated with an iron hand all the affairs of society; that every year enlarged its power, every move increased its dominion; that the men and women who dared to even question the divinity of that institution were ostracized, were persecuted, were villified—aye, were hanged.

"But the great clock in the chamber of the Omnipitent never stands still. It ticked away the years as it had ticked away the centuries. Finally it struck the hour and the world heard the tread of a million armed men, and slavery vanished from America forever."

Note the parallel: Today the syndicate rules at the White House and makes laws at the capitol; courts of justice are its ministers; senators and legislators are its lackeys. It controls the preacher in his pulpit, the professor in his lecture room, the editor in his sanctum; it swaggers in the drawing room; it rules at the clubs; it dominates with a rod of iron the affairs of society. Every year enlarges its power, and the men and women who protest against the crimes that are being committed by organized greed in this country—who talk of protecting the American people—are ostracized and villified, are hounded and imprisoned.

It seems madness to even question the divinity of the American Syndicate. But, my friends, the great clock is still ticking—still ticking. Soon it will again strike the hour, and the world will see not one million but ten million freemen rise up, armed not with muskets, but with freemen's ballots, and the sway of the syndicate will vanish from American forever."

These prophetic words are to be fulfilled through the organization, education and co-operation of our millions of farmers.

LET THE WATCHWORD BE ORGANIZATION.

Editor of the Equity Union Exchange:—Long ere this article was written it had been a foregone conclusion that the farmer and producer must find a means of protection for the fruits of his labor. For every article which he produces he must take the price the other man may set, and pay the price asked for the goods he has raised, when they are offered in manufactured form.

Friend farmer, this is not fair to ourselves. So it is to you I make the plea to organize into a "combine", like the Equity Union, which has the best plan for protecting our interests that we have found yet.

The farmers may be, and they certainly should be, the most independent class of people on earth, and yet they are dependent on the other man for advise and price on their products, and it is a shame, brother farmer, to allow this to go any farther. How shall we meet this proposition? Every one is to put his shoulder to the wheel, and in a conservative manner push and push hard. At primaries and general elections we select city bred men to represent us in the legislature, to make and enact laws, to uphold trusts and combines for the purpose of frustrating our own plans, putting middlemen into the field, to pay them a per cent to sell us our goods back to us. A grand work, is it not, my brother farmer?

We need not be too optistimistic or pessimistic in this matter, but conservative, and use good common sense and organize into Local and National Unions.

Every other occupation is organized, and all have made a grand success and have gained cognizance before the great business world of today. As our Saviour said: "Awake, thou that sleepeth, arise from the dead and I shall give thee light." We realize that we have been and are back numbers as far as our own defense has been concerned.

We do not expect to have every thing our own way, neither do we want it in this form. "Brother with brother, heart to heart and hand in hand," should be our motto, with love to all and malice toward none. Not to make consumers pay more, but less, to do away with the middle men's profits.

We labor with our brain and muscle, why should not they, the middle men? We can build our own mills, our own elevators, ship our stock and produce, and buy our goods direct by the car load. Shall we do it? Answer, "Yes." Right about face, and now to action. Who will be ready to do the work in his neighborhood, working faithfully winter and summer? The time has come for action. Examine yourself and see if you are not the man.

Regent, N. D.   F. M. CASS.

THE EQUITY UNION THE GREAT ORGANIZING FORCE.

The Farmers' Equity Union is being firmly planted in Ohio, Indiana, Illinois, Missouri, Oklahoma, Kansas, Colorado, Nebraska, South and North Dakota, Minnesota and in Washington on the Pacific coast. When strongly organized it will prevent gluts of central markets, which now cause farmers to lose millions of dollars on their largest crops of finest quality. When their is sufficient co-operation by our Exchanges, the price of farm machinery will be reduced fifty per cent, coal will be reduced $1 or $2 per ton and flour more than thirty cents per hundred pounds. Both consumers and producers will be benefited. Hundreds of Equity Exchanges must be organized at the best country points.

From five to ten thousand dollars capital is raised at each shipping point where we have an Exchange. The stockholders control this capital and their own business. The National Union unites all of the Exchanges, leads them to co-operate and carries on a continual campaign of organization and education. This education is absolutely necessary to success. A co-operative paper reaches every member's home weekly. We tie every member to that paper for four years and ask every one to write for it weekly. Every local Union ought to have a correspondent. When two hundred farmers are organized and educated at one place we organize a stock company and establish an Exchange. The shares are twenty-five dollars each, and the limit four shares. The manager is properly bonded and his books are audited regularly. Each Exchange has a cut off in June and January and an expert bookkeeper audits the books for six months. Our Union furnishes a uniform set of books to all of our Exchanges, and sends a man to instruct our bookkeepers and managers fully how to run the business successfully. The manager is required to buy and sell on a safe margin. A large volume of trade from two or three hundred good farmers reduces the cost of handling and insures the success of the business. No danger of assessments on shares here.

Out of the gross earnings are taken expenses, national dues and five per cent dividends on stock subscribed. The directors can never declare over five per cent dividends on the stock subscribed. All over this is net earnings, and is figured as a per cent of the entire business, and is prorated back to stockholders according to the amount of patronage furnished. We are sure that if two or three hundred farmers will unite their trade for grain, wool, stock, produce, flour, feed, fencing, salt, coal and all farm machinery they can pay themselves five per cent for the use of the capital invested, and ten per cent of all business furnished can be paid back for patronage. Then five hundred dollars worth of patronage will earn fifty dollars and is paid back to the farmer who furnishes the patronage. One thousand dollars worth of patronage will earn one hundred dollars worth, etc., etc.

Every patron who is a stockholder gets back all that he earns by his patronage. Only patrons can draw out the earnings of the company, as patronage makes the earnings. This plan is just and holds the farmers together when they understand it. A small per cent for capital invested, and as large an amount as possible paid back each year for patronage, will bind a million farmers together in a Union that will be powerful and beneficient. A safe margin, a large volume of trade, a small per cent on money invested and cash paid for patronage will guarantee the success of every Exchange and bind ninety per cent of our farmers together.

Each members' produce is handled at actual cost, and a large volume of trade reduces the cost of handling. Non-members are paid the same price as members for their crops and buy as cheaply but do not share in the profits. We pay all of the earnings of the company to patrons who are stockholders. They must be both patrons and stockholders to get their share of the earnings. Non-members will soon see the difference and come into the Union. They cannot afford to stay out. This plan will unite one million farmers.

The entrance fee is three dollars each which pays for the paper for four years. Each stockholder is equal in vote to any other, no one being allowed to own less nor more than four shares. To all members having less than four shares, we prorate in shares for their patronage till they have the limit, four shares. A Union of two hundred farmers will finally have twenty thousand dollars invested and banked. Their financial standing will be twenty thousand dollars in the business world, which gives them a wonderful power.

All of our Exchanges must co-operate as much as possible in buying and selling. We must buy all of our wagons, drills, binders, etc. from the same factories. Each members' produce will have an advantage in the markets. Union farmers and union labor members must give his produce the preference. We will buy the entire output of mines, mills, and factories, at greatly reduced prices. In fact it will not be long before we are grinding our wheat in Equity Union mills.

We will work out an economic plan for the distribution of our produce direct to the consumers.

We are organizing a fine market for apples in Minnesota, the

Dakotas and western Nebraska and Kansas which must be supplied by our Unions in Washington, Missouri, Illinois and Indiana. We do not want a million bushels of apples to rot on the ground while the price in the Dakotas and western Nebraska and Kansas is so high that common people cannot afford to buy them. Our plan is just and practical. We ask every farmer who reads this book to join our Union and take at least one share in an Equity Exchange.

THE FIRST SATURDAY, EQUITY UNION DAY.

We are endeavoring to organize and educate one million farmers and their families to quit work and lay aside all other business the first Saturday of every month, and attend an Equity Union meeting. The entire family should go if possible. We want two Sundays every week for Equity Union farmers. One must be given to organization for economic freedom, the other to spiritual power and uplift of humanity. The power and benefit of these meetings cannot be measured.

An opportunity would be afforded for all the farmers and their families to become better acquainted and more fraternal. The exchange of friendly greetings, practically ideas about farming and co-operation and the transaction of all necessary business will all conduce to the success of the Equity Union movement.

If a million farmers were meeting regularly the first Saturday of every month, they themselves would be wonderfully astonished at the result. Steady, equitable prices would be maintained on all farm produce. The price of all food products would be reduced twenty-five per cent. The unnecessary profit takers would be out of a job and millions of dollars which they now TAKE would go into the country to the wealth-makers for good roads, good soil, good homes, good schools and churches.

Farmers, we must meet together. We must work for big meetings. Money enough must be placed and kept in each local Union treasury for stamps and stationery and all expenses, so the secretary can carry on all necessary correspondence promptly and efficiently. The secretary should also have sufficient funds in the treasury to send two hundred farmers a printed card each month notifying and inviting them to the meeting. The local Union which does this regularly every month, a few days before the monthly meeting, is sure to grow to a membership of more than two hundred members. If the farmers meet they will become interested and their interest will be kept up. They will pay all dues necessary, and thus financial support will besecured.

Meetings and money are absolutely indispensible to a good successful Union. A successful union means so much to farmers that we can not afford to neglect our meetings the first Saturday of every month.

The first Saturday of December is especially important. Local

Union officers are elected for one year. A delegate to the national meeting must be elected, and changes in the constitution may be made by local unions providing they do not conflict with the National Union by-laws.

Give us fifty or one hundred thousand farmers meeting regularly the first Saturday of every month and we will get better and better deals every year on twine, fertilizer, tile, salt, apples, potatoes, wagons and all farm machinery. Each local union then can canvass its territory thoroughly every month and ascertain how many farmers will buy wagons, binders, headers, mowers, gasoline engines, automobiles and all farm machinery. New members would come in by the thousands from every state where Equity Union is started.

THE CAUSE OF HEAVY LOSSES TO FARMERS.

A Birmingham banker addressing the cotton growers convention in September, 1911, said: "It is certainly false economy to rush your products pelmel to market to be placed upon the bargain counter and bought at what ever price the buyer chooses to dictate. Something is wrong when the boll-weevil and the worms are called a blessing because they reduce the size of the crop. Something has gone awry when a crop of eleven million bales will sell for more money than a crop of fourteen million bales."

Every farmer must have this condition of things rubbed into him until he will act on a proposition to organize and market co-operatively. The fact as stated by the banker in regard to cotton is true of grain, stock, apples, potatoes, cabbage, onions and all farm crops. The larger the crop the less money farmers receive for it. We sell our largest crops of finest quality below cost of production. Others make millions of dollars clear profits from the crops while the farmers, who produce the wealth and have it all in their hands once each year, actually sell at a low, unjust price.

The cause for this is not hard to find. The farmers are an unorganized mob, and the principle buyers in all of our leading markets are thoroughly organized. The farmers compete while the buyers co-operate. The farmers rush their crops to market "to be placed upon the bargain counter" and bought at whatever price the buyers choose to dictate." The organized buyers handle the farmers one at a time, just as the old man handled the bundle of sticks. The individual farmer has no power to defend himself against the dumping system. His only hope is a National Union of farmers, which will break up the dumping system and work out an intelligent system of distribution of all of our valuable crops. When a pack of wolves got into a pasture over in Missouri and began to circle around a drove of mules, the mules slowly backed away toward the center of the pasture, where they got together, but as they had backed together their heels were together and they began kicking one another and soon two of them were down flat on the ground and the wolves had a fine feast on mule meat. However, a long headed old donkey in the next pasture, observing what had occurred, at once organized the mules in his pasture by getting them to all put their heads together, instead of their heels, and when the pack of wolves came, they kicked wolves instead of kicking each other. We hope farmers will show as much sense as the last bunch of Missouri mules.

THE PRACTICAL AND SELF-SUPPORTING.

The self-supporting feature of the Equity Union appeals to every farmer when he understands our practical plan of co-operation. Neither the Union nor our Equity Exchanges cost our members one cent. Our entrance fee of three dollars to the Union brings to each member the Equity Union Exchange for four years, the best co-operators paper in our country. This gives him full value received, but we pay him back the three dollars in cash for his patronage.

Each member must take one share of twenty-five dollars in the Exchange to get the full benefit of our co-operation, but when thoroughly organized and our plan is fully carried out, we are able to pay him back the twenty-five dollars in cash for his patronage, and to donate to him four shares in the Exchange, by giving him the profit on his patronage, which he now gives to the middle man. There is no visionary scheme about this. It is simply a practical plan of golden rule co-operation, which is being carried out successfully at about one hundred good markets, where farmers are wise enough and fraternal enough to organize and apply these righteous principles.

The Tamarack Co-operative Association in Michigan has operated successfully for twenty years. The manager's report shows $64,610 as the capital stock paid in. The twentieth annual dividend of $104,821.60 was declared in February, 1911. This was one hundred sixty-two per cent on the investment, but the Michiganders are true golden rule co-operators, and would not think of declaring this enormous profit on the stock subscribed. They are aware that big dividends on dollars invested, rob wealth makers and make millionaires of wealth takers. This dividend was prorated back to stockholders according to patronage.

About two thousand families owned the stock and furnished the patronage. The average family purchased about four hundred thirty dollars worth of goods and received a rebate of twelve per cent, almost one-eighth of the purchase price or $51.60 per family, in addition to the interest on the money invested in a share of stock. Since they started in business this Union has done a business of $8,113,911.85 and has returned patronage dividends to the extent of $938,033.67 to its members. In twenty years these co-operators have saved nearly $1,000,000. We have at Tamarack, Michigan, a practical demonstration of the Equity Union co-operative idea.

The grangers of Johnson county, Kansas, have carried out practical co-operation very successfully for over thirty years. The

Jackson county co-operative company, Lakefield, Minnesota did a business of $139,230.86 in 1910, and gained $12,700.21 and prorated back sixty dollars to each family for patronage, after a dividend of six per cent had been paid on capital stock. We are sure that when the Equity Union plan is fully carried out by our one hundred Equity Exchanges, that five per cent can be paid for the money invested and ten per cent of all business furnished can be prorated back for patronage. Then for every one thousand dollars of business furnished by a member, he will receive one hundred dollars as a patronage dividend.

The members of our 150 big unions are being educated on the principles of true blue golden rule co-operation, and they will demonstrate more and more from year to year, that our Union is self-supporting and that neither the Union or our Exchange will cost one cent, when the Equity Union plan is fully carried.

We do not believe it right nor wise to prorate to non-members. They must unite with our Union if they want benefits. They need the same protection as we do, and should be with us and not against us in this great movement for the uplift of humanity. The multitudes are very willing for the few to carry the load for all. We are trying the dollars and cents persuasion to unite the farmers. We are sure it will work. Cash prorated for patronage talks. It beats lecturers and literature.

If every farmer will give up his suspicions, throw away his doubts and fears and come into our Union, go to every meeting, pay his one dollar national dues each year, read the Equity Union Exchange, and work for members and stockholders in the Equity Exchange, we will in a few years work out a wonderful economic system of distribution which will "astonish the natives," and be worth hundreds of millions of dollars to us as producers as well as to the consumers.

Farmers are standing the loss of millions of dollars each year from hog cholera, chinch bugs, hessian fly, boll-weevil and other pests and they alone must stand the great loss. They ought to unite and prevent the low, unjust prices which are sure to come every time there is a good crop of hogs, grain or cotton. The business man always figures in the risk as well as labor, expense and time, and then adds on a certain per cent of profit. Farmers must organize and carry out intelligent business principles of co-operation both in buying and selling. The honest, intelligent distribution of our products direct to the consumers is a great problem. The present system holds the farmers' prices down to the lowest notch, and the consumers' prices up to the highest pinnacle.

The Equity Union is reversing this robber system. We establish a market at the best shipping points which will give the farmer a just price every time and sell to him at a reasonable price. Where this is done you do not see farmers who sell cow peas receive one dollar a bushel while farmers who buy them for seed pay three and four dollars a bushel. You do not see good apples sell for thirty cents a bushel while consumers in many localities pay three dollars a bushel. Onion growers will not be compelled to sell good onions for forty cents a bushel while consumers pay three dollars a bushel. Seventy per cent of the winter wheat left the farmers' hands in 1911 at eighty cents per bushel, and the speculators held it and received one dollar and twenty cents per bushel before the next crop came.

Every farmer ought to assist as much as possible in establishing a market in his town that will pay him a just price for his produce and sell to him all that he buys as cheaply as possible. A large majority of our farmers need that kind of a market very badly. Farmers must remember that God helps those who help themselves. We must work hard to form our own Union—just as railroaders, miners, mechanics and factory workers have organized theirs, and then we will work out an intelligent system of marketing. Wage workers have been driven to it, but can never be fully successful until farmers work out a system of distribution in the interests of both producers and consumers. Wage workers may strike and secure higher wages but so long as trusts fix the prices on the necessaries of life and raise them every time the wage workers are successful in securing higher wages, we cannot see where the wage workers have really bettered their condition.

THE NEW LINE OF EQUITY ELEVATORS.

Thousands of old Line elevators have been built in the grain sections of our country during the past forty years by a set of speculators for the purpose of handling the farmers' grain at outrageous profits. In some towns ten elevators have been erected to handle the grain of bumper crops which were all thrown on the market at threshing time. The farmers were compelled to pay for all of these elevators, which stand idle three-fourths of the year. They pay for them again and again, and yet the speculators still own them and continue the same robber system from year to year. The farmers have been robbed of millions of dollars by this infamous profit system.

At New England, North Dakota there are six big robber concerns and one Equity Exchange. The farmers have marketed as high as two million bushels of grain in one year at New England and the profit takers would have the cheek and nerve to take out of each such crop at least one hundred thousand dollars if there were no Equity Exchange there. These vultures are always ready to pounce upon their victims whenever a crop is raised. The farmers have all the labor, expense and risk of production and marketing, but when they finally succeed in raising a bumper crop of fine quality, after two or three failures, they haul it in to the elvetators, and have it priced, weighed, graded and docked according to the mercy (?) of Mr. Capitalist.

Millions of farmers and their hard working families are at the mercy of cold blooded sharks, who plunder them right and left, simply because the farmers are not organized for self-protection.

If any intelligent, honest man will study the conditions on the frontier as I have done for ten years, he will be fully convinced that their is no exaggeration in these statements. I have spent the last ten years organizing in Minnesota, North Dakota, South Dakota, Kansas, Oklahoma, Colorado and Nebraska, and have visited in the homesteader's sod-house homes and learned their condition from first hands. They are the most courageous, enterprising people in one way, to be found in our country. They braved the blizzards, they tackled the wild lands, and encountered all the privations and hardships of a frontier life in order to have homes for their families. Many of them are poor people contending with failures of crops and the money sharks at the same time. The farmers have organized stock companies and erected elevators for self-protection in some localities. These have brought down the margins of the speculators and benefited all of the farmers to some extent. But most of these companies are still run on the capitalistic plan of making big dividends on the stock subscribed. The results are not satisfactory to a majority of the farmers. A few rich farmers or business men of the town get control of the stock and skin the rest of the farmers for big profits on their dollars. They adhere to the old Line principle of big per cents on the almighty dollar.

The Equity Union will educate the farmers in some of these communities to change their plan and adopt the co-operative plan of paying only 5 per cent dividends on stock subscribed and prorating back as large a per cent as possible for patronage. The Equity Union idea is right. Patronage earns dollars and not capital invested. The working capital is furnished by the patrons instead of Mr. Capitalist, and hence the patrons get the earnings of our elevators according to patronage, instead of Mr. Capitalist, who often proves himself a shark, cold-blooded and merciless.

So the Farmers' Equity Union is starting a new line of elevators from North Dakota to Texas, and from Ohio to Montana, that will revolutionize the whole system and change the methods of doing business among the millions of farmers.

We often start them on a small scale and develop the business as their capital increases, and the farmers are educated to co-operate. One danger is haste in branching out without sufficient capital or education of the people. Do not try to go too fast.

A few good centers are being worked up in Ohio, Indiana, Illinois, Missouri, Kansas, Oklahoma, Colorado, Nebraska, South Dakota and North Dakota. As these Exchanges are firmly established and made successful, they are educators to other shipping points near them. The circle must be widened year after year around Fort Wayne, Ind., Greenville, Ill., Aberdeen, S. Dak., Kansas City, Kans., and Oxford, Neb., until they meet and form one great solid National Union of farmers. Then each shipping point will have one elevator instead of from 3 to 10 elevators. The farmers will center a large volume of trade together, be their own capitalists and run their own business co-operatively, without the assistance of grafters, profit-takers, or capitalists.

The new line of elevators or exchanges are all combined or linked together under one National Head in the Equity Union. We are a great co-operative, industrial Union, and the champion of the industrial rights of every man, woman and child on the farms.

The new line of elevators are beginning to co-operate in hundreds of ways for the interests of each other. There is no rivalry nor competition between them but constant co-operation. The Exchanges in the Dakotas, Nebraska and other parts where no apples can be raised, buy direct from our Equity Union members or Exchanges in Washington state or Indiana, Illinois or Missouri. The apples are usually loaded in bulk in good shape in refrigerator cars. We want Equity Union apples, potatoes, corn, oats, hay, and stock shipped directly from one Equity Exchange to another, where it is possible, so as to cut out as many toll-gates as possible. We are now in our infancy and limited in means, but this movement has a wonderful future, and promises great blessings to the people, when they are organized in the Equity Union, and educated to carry out true blue golden rule co-operation.

The farmers will finally see the great difference between the old line and the new line of elevators or Equity Exchangs. They will see in many localities a difference of five cents a bushel on grain or a difference of fifty thousand dollars on each million bushels of grain marketed, in favor of the farmers. When the new line is as strong as the old line has been, gluts of central markets will cease and there will be a difference of twenty cents a bushel in the price, which our poorest farmers now lose on the largest crops of finest quality, and give up to speculators.

The Equity Exchanges are introducing the Rochdale system of buying and selling and solving the question of the high cost of living to some extent. Our Exchanges are run on just principles. We teach the Saviour's golden rule. The man is the unit and not the dollar. We are for men. The poor member is protected and benefited as well as the rich. Every farmer who will join can have the benefit of co-operation. We hold the door wide open and persuade every one to come in. We show them the benefits of co-operation in dollars and cents, so they can count it. The demonstration in cash is a wonderful educator. If grain is bought on the present margins and only five per cent dividends paid on the stock subscribed, at least five per cent will be made on the business, which is prorated back to stockholders for patronage, so the patrons who are members will receive from ten to one hundred dollars each year for patronage.

The principle is just and the plan practical. It will unite ninety per cent of the farmers at each good shipping point. Thousands of links can be made for this Equity chain on this plan. The cooperation of the links will reduce the price of farm machinery fifty per cent. Fair prices on all farm crops can be established and maintained from month to month, and our prices be as steady and sure as the price of anything we buy. The success of this movement depends largely on putting a good weekly co-operative paper into the home of every member, which teaches co-operative buying and selling. Many of our agricultural papers are owned or controled by the trusts, and hence refuse to open their columns to the discussion of this question. The trusts buy them with profitable adds.

Every member who joins this Union must be a subscriber to our official paper, the Equity Union Exchange, for four years. Its columns are open for the education of farmers for their own interest and not for the trusts. We want our members to write all encouraging news. Get every farmer in your neighborhood to subscribe for our paper and he will find it to be a clean paper and the true blue champion of the farmers' cause. It deserves the support of every farmer in the United States.

CO-OPERATION VS. COMPETITION

That competition is the life of trade, is a doctrine which is being discarded in the business world. We all know General Sherman's expression about war—Competition is war. It is one of the causes of the high cost of living. In the price of farm machinery is included the price of unnecessary cost of advertising and of unnecessary traveling men to the extent of hundreds of millions of dollars, all of which is paid by farmers. Nearly everything which we have as consumers purchased, has this unnecessary cost attached to it.

The factories, mills and mines fight each other in the competitive world, for our trade at enormous expense, but they combine sufficiently to make us pay the bill. The heavy cost of advertising and of traveling salesmen is added to the selling price and farmers pay it.

The cost of manufacturing a machine is only a trifle when compared with the cost of selling it. This high cost of selling is caused by competition. There is continual war between business concerns for our patronage and some body must pay the heavy cost of war.

There are too many mills, mines and factories, and too many wholesalers, jobbers and retailers all fighting each other at heavy expense for our trade and spending millions of dollars to secure it. The expense of getting the farmers' trade in the competitive world, is in many cases, three or four times the cost of manufacturing the machines. Self-binders sell for $150, while the material and labor costs less than $50. Wagons, which cost less than $25 for material and labor, sell from $75 up to $100 each, and so with all farm machinery.

The competitive system and trust system are responsible for high prices which farmers pay. Competition among farmers in selling their valuable crops often reduces their prices below cost of production.

Millions of dollars are lost to farmers in this way every year. We pay the high cost of competition among manufacturers, and then break down our own prices in all the leading markets, and all over the country by competing in the sale of our products. No other class is hurt so badly as the farmers of this the greatest agricutural country in the world.

The Farmers' Equity Union has a co-operative plan of buying and selling which will eliminate competition with its contsant struggle and expensive warfare, and save millions of dollars to farmers when fully carried out by them. We are urging farmers everywhere to study the Equity Text Book and become thoroughly acquainted with the plan and principles of this new organization which is proving a grand success at more than one hundred large markets. When this plan is fully carried out, it means economic freedom to all the people.

Farmers, our defense is in golden rule co-operation. We have the weapons of defense in our own hands but refuse to unite and strike the blow for freedom. Farmers are sleeping by their loaded guns, and must be aroused and pesruaded to use the effective weapon—co-operation. The mere discussion of this question fills every grafter, speculator, extortioner and price-hammerer with rage and alarm. They know our power when united and fear the righteous indignation of farmers, if they ever should be awakened and become true blue golden rule co-operators. There is no need whatever for farmers to submit to any wrong. They have all power when united and educated to co-operate.

The mission of the Equity Union is to organize farmers and persuade them to substitute co-operation for competition. By competition we play right into the hands of the trusts and speculators and sell our largest crops of finest quality for low unjust prices. When a thousand Equity Exchanges are linked together in one great chain, we will show the great value of co-operative buying and selling on a national scale.

The Exchanges must be built up on the Equity Union plan, till they have financial standing in the business world to the extent, of ten or fifteen thousand dollars each. This is done by centering a large volume of trade together on a safe margin, and adding all profits to the capital of the Exchange, instead of giving it to Mr. Millionaire-Profit-Taker. When the financial standing of each Exchange is established in the business world, the National Head can do business for all of them, collectively, with any factory in our country, or any other country. An order for fifty wagons from each Exchange will be honored and accepted by any factory, so that the order for fifty thousand wagons can be centered with one factory one year in advance. Then this factory would be taken out of the great war of competition, with its expense, worry and risk, over into our co-operative world of peace, prosperity and plenty. Union labor must be employed in this factory. It is enabled to run steadily the entire year, giving steady employment to the men as well as good wages. There is no risk of sale, no fear of competition; and no alarm when the tariff is discussed by congress. The wagons are all sold at a fair profit to the manufacturer. We cut down all expense of advertising and traveling men and all risk of sale. The entire output is sold when he begins to make them. We go to the factory with out patronage organized. The expensive war of competition is prevented by co-operation of the Equity Exchanges.

On this co-operative plan wagons, self-binders and all farm machinery can be reduced fully one-half in price, when enough farmers become golden rule co-operators.

The entire output of factories, mines and mills will be taken by the great National Union of farmers. Unnecessary mills, mines and factories will be eliminated. Only those needed will be retained. We must widen the circles more and more around the centers where we have made a good start. One link will call for another. The demonstration at one point will educate the farmers at the next station. The Exchanges in the apple section will find a safe, profitable system of distribution. The Exchanges needing apples buy direct from them. The demand for Equity Union apples grows stronger from year to year. Many of our Exchanges are asking for Equity Union potatoes and cabbage. They want no "scab" potatoes nor apples, but Union produce. The farmer who refuses to unite is an enemy to himself, his family, his community and the entire farming fraternity.

The success of this movement is so important to all farmers, that no man who loves humanity will hold it back by indifference, negligence or opposition. The worst foes are the Can't-do-its and Won't-Sticks. But the courageous, public-spirited, generous-hearted workers are pushing ahead in spite of all opposition, and are sure to win out because they are right. We must have the three M's—meetings, members and money. Every member must work for one more. Choose up, have a contest for three months for new members. Let each member who is present at a meeting count five for his side, and each new member brought in with the three dollar fees count twenty and each new stockholder twenty-five, and the side at the end of the contest, having the largest number will be the winners. The side losing must furnish the chicken dinner or oyster supper. Some local unions have doubled their membership in this way. The annual dues must be collected for the National Union in each November. Read the by-laws on this subject. Very little progress can be made without working members, attractive meetings and liberal, financial support. One million farmers can hold on penuriously to that dollar a year dues and kill a grand movement worth hundreds of millions of dollars to them annually, but if each one would make it his religious duty to put that dollar a year in the National Union treasury, and then see that it was used right, we would have hundreds of the best organizers and educators in the field constantly and in a short time co-operation would take the place of competition in our business world to the great advantage of the millions of farmers and their families.

As Equity Exchanges are established and operated, financial support ought to come to local unions and the National Union, so that a great national educational campaign can be inaugurated and carried on continually. This is absolutely necessary to the success of the Farmers' Equity Union. Farmers will not organize themselves nationally. Education is a slow process, but it brings sure, permanent blessings to the people.

The overthrow of the costly, warring, competitive system and the complete establishment of true blue golden rule co-operation can only come as the farmers are organized and educated to be more intelligent, moral and fraternal. The same movement which brings economic freedom will bring blessings far more important to the people. The general uplift of humanity will be the result. Economic freedom and fraternalism must go hand in hand. Fraternalism and co-operation are twin sisters.

A divided people are a weak people, and the easy prey for combined grafters and speculators. A united people are invincible against every foe. Farmers, let us unite, be fraternal, and build a great National Union of farmers that will overthrow the competitive system, which costs the people untold millions of dollars annually.

PROGRESSIVE FARMERS

The most progressive people in the world are found in the western part of the United States, in that section of our great country which was marked "American Desert," in our geography when we went to school in the little red school house on the hill. There are very few standpatters among them. In starting, this new Equity Union spent six months of hard labor organizing in Indiana, Illinois and Missouri, and only started four local unions in the six months. But the first four months spent in the Dakotas resulted in seventeen strong local unions, some of which have over 200 members at this writing. Six months work in Kansas started twenty seven local unions, nearly all of which have organized into Equity Exchanges.

The energetic, open-hearted, courageous farmers of the west are demonstrating the practibility of our plan of golden rule co-operation. They are open to conviction, ready for investigation, and will adopt co-operation when ever shown a practical plan. They are the most generous hearted, public spirited people on earth. They never stand in their own light one moment. They are capable of taking care of themselves and soon learn to look after their own business in a business way.

We have often wished that our standpat friends of Illinois and Missouri, and our progressive friends, too, could have seen the eighty nine farmers swarm around the secretaries at Bowman, N. D., when we organized there in July 1914. At St. Francis, Kansas, in September 1913, sixty-seven farmers joined the Union and paid their fees at the first meeting. At Mott, N. D., nearly three hundred farmers are in line for Equity Union, and own two large elevators and a large lumber yard; they paid back to themselves nearly fifteen thousand dollars in profits in 1913 and 1914. Liberal, Kansas, prorated over seven thousand dollars June 21st, as a result of one year of Equity Union co-operation. The very atmosphere in the west seems charged with a spirit of freedom. Show them when and where to strike the blow for liberty, and they will strike it promptly and effectively. The west is always teaching the east lessons of progression. While this is true of the progressive western people, we are sure that Equity Union will grow stronger wherever it is planted in Missouri, Illinois, Indiana and Ohio among the most progressive farmers.

WHO SHOULD HAVE THE PROFITS?

The profit-takers are abroad in the land. They are entirely too numerous for the interests of the millions of farmers. They are found among the manufacturers, rail roaders, bankers, speculators, jobbers, wholesalers and retailers. The business world it full of them. Every country town has them by the score. The central markets are overrun with them. There are large and small profit-takers. They all live by profit taking. Many of them live at ease and in luxury by this business. They fare sumptuously every day. They are arrayed in purple and linen. They toil not, neither do they spin, and yet Solomon in all his glory was not arrayed like some of them. Many of them pile millions of dollars in their unholy coffers annually

When our products from the farms reach the consumers, they have often passed through the hands of six or seven profit-takers, and the consumers pay four or five times the farmers' price.

Enormous profits rob the farmers and their market—the consumers, and make thousands of millionaires. Outrageous profits have made the money-kings of our country and are adding more and more to their power to oppress the people and rob them of their economic liberty.

The profit-system is the prolific cause of pauperism and of the straightened circumstances of millions of honest and industrious farmers and their families. This nefarious system is often the cause of low prices to farmers and high prices to consumers. The profits of many large business concerns have run so high in many instances, that they resort to watering their stock to hide the enormous dividends. Co-operation abhors this unholy system.

The Farmers' Equity Union teaches golden rule co-operation—Do unto others as you would have them do to you. Quit trying to do the other fellow. Love your fellowman. Give him a square deal. Be fraternal. Down your own selfishness. Be for the other fellow and not so much for yourself. It is right. It will pay.

We are uniting the farmers and educating them to run their own business without Mr. Profit-taker. Farmers are gradually getting it into their heads that it pays to be fraternal and co-operate in business. They are learning to handle their own business co-operatively without profit. They are entering the wedge that will destroy the infamous profit-system, and give them the wealth they produce. They are awakening to the fact, that there is a wonderful protection in Union and co-operation. It is their only defense against the robber profit-system.

When a farmer drives up to an Equity Exchange with a load of grain, stock or cotton, he is not worried about grading, pricing, or big margins. He knows also that his produce is handled at actual cost. No profit-taker stands there to lower the grade and price as much as possible, in order that his profit may be large. When he must have seed wheat, flax or oats, he is not held up by Mr. Profit-Taker to the tune of fifty per cent. He is in a position to protect himself every time; if he is a member of the Equity Union all profit, if any, is counted back to him in cash at the end of each year.

All of the earnings of each Exchange are counted back into the pockets of patrons, instead of the pocket of a millionaire-robber-profit-taker. Patronage makes earnings and not Mr. Profit-taker's dollars. If he wants to eat he will have to work, as he can no longer take what others make. He will have to obey the divine command, "In the sweat of thy face shalt thou eat bread," just as farmers do.

Farmers, join the Union and be golden rule co-operators. It is our only hope. Come under the banner of the Equity Union. There is protection, prosperity and peace under its folds. Investigate our principles thoroughly. They will bear investigation. They are just and our plan is practical and easily carried out when enough join the Union. When we are educated to handle our own produce we will branch out and handle all we buy and sell at actual cost—without profit. But we urge our progressive members not to try to go too fast in business. Go slow and educate—educate—and build up your capital by safe margins and by prorating shares for patronage.

Get at least ten thousand dollars capital in each Exchange. One hundred members at one hundred dollars each will give this amount of capital. Never have a sinking fund nor a surplus fund. Work for more stockholders and patrons. Unite more and more farmers, and remember that every new stockholder means one hundred dollars more in the capital. We want one million farmers united against the profit-takers. This will make sure work of that system. We must overthrow the profit-system and quit making millionaires. One hundred good farmers can be united at any good shipping point and the shares of each farmer built up by his patronage till he has one hundred dollars of the capital. This will give the Exchange ten thousand dollars capital. If more capital is needed, increase the limit of shares to five and you will add twenty-five hundred dollars to the capital. Buy and sell on a safe margin, pay back all profits for partonage and in this way build up your capital.

When a million farmers are united in the Equity Union, profit-takers and millionaires will not be so numerous, but beautiful homes, schools and churches in the country will be more plentiful, and the President of the United States will not have to appoint a Country Life Commission to see what is the matter with the farmers' condition. We are well able farmers, to run our own business, without all these toll gates with the profit-taker at each gate.

When united we will make the farmers' business the best business in the country, and make conditions that will attract our boys and girls instead of driving them to the cities.

We will act on the principle that all profits on business should go to those who furnish the business. When this principle is carried out by every farmers' elevator company in our country, we will have one million farmers united in a short time.

SOLUTION OF THE TRUST PROBLEM

Ex-president Taft in his address to the G. A. R. at Rochester, N. Y. August 23rd, 1911, said: "So long as we retain in this country a God-fearing, sober, intelligent people, we can count in the long run upon their working out safely and sanely the problems set before them, no matter how often they have been defeated in their purposes by the temporary success of undue and corrupt influences of concentrated wealth."

Two great facts are stated here by the ex-president worthy of the thoughtful consideration of every good citizen. Concentrated wealth is the curse of our nation. The system by which the wealth of the toiling masses is concentrated in the hands of the few, holds a large majority of our people in this so called free country, in industrial slavery, some in a greater degree than others. The beef-trust owned by two or three men controls nearly all of the packing plants in Chicago, St. Louis, Kansas City, St. Joseph, Omaha, Sioux City, Sioux Falls and St. Paul. They control and manipulate the farmers' markets for hogs and cattle so as to make feeding stock very much like gambling on the Board of Trade. They also fix an arbitrary price on consumers, so there is no hazard whatever about their prices and business, no doubt whatever about millions of dollars rolling into their unholy coffers annually from the hard earnings from the toiling masses.

Farmers, listen! Shall only prices and profits of the trusts be sure in this country? Shall the wealth takers be allowed to combine and be the only people whose profits and prices are sure every year whether crops fail or not?

The beef trust not only controls nearly all of our meat products, but they own the tanneries, control the leather, own the machinery used in our shoe factories and have recently purchased a large shoe factory in St. Louis. They now propose to fix the price the people shall pay for meat, leather, harness and shoes. It seems that the price of everything we buy and sell on our farms is finally to be fixed arbitrarily, without regard to justice or injustice, by a few greedy, selfish millionaire magnates, whose greed for gold knows no bounds and crushes out of their bosoms all regard and love for their fellowmen, and makes them monsters ready and willing to crucify humanity on the cross of gold.

This beef trust is only one of the many trusts all of which are run on about the same principle of the greatest good to the smallest number. All the trusts are intermingled together. Each magnate holds some stock in all of the trusts, and has a directorship some times in more than a dozen organizations. All seek more and more to concentrate the billions of dollars of wealth produced each year by the wealth makers in the hands of the few. The combinations control our markets and fix prices arbitrarily.

They control our state legislatures and our national congress, and even dare to threaten the country with a panic when the National Administration attempts to make them obey the laws as common people do. The President, as well as every other intelligent citizen, is aware of the fact, that "undue and corrupt influences," are continually at work in the political world by the millionaire wealth takers. In coming elections every citizen should know for whom he is voting. Not what party label does he bear but what principles does he stand for, and is he an honest, sincere man working for the interests of the people, or is he a stand patter for special interests. Farmers, let us require every candidate to define clearly his position on referendum, initiative, recall, direct primaries, the parcels post, the extension of postal savings banks and extension and enlargement of their powers till they will receive any amount from a citizen at two per cent interest and loan any amount to a citizen on good security at three per cent interest. Then our banking system will be in safe hands, and the power of the "Money Truts", will be broken and panics in the future will be impossible. Big per cents, big dividends and big profits on dollars are the prolific causes of millionaires. Every candidate for the legislature ought to be pledged to work for the Kansas Co-operative Law.

In all states we have corporation laws, but in most of the states we have no law recognizing a purely co-operative company and protecting our co-operative principles. We need all these to break the power of political bosses, corrupt legislatures, and to prevent "undue and corrupt influences of concentrated wealth." We believe with ex-president Taft, that the number of God-fearing, intelligent and sober people in our country is increasing and that they can be depended on finally to work out safely and sanely the economic and political problems before them, in spite of the temporary success of "undue and corrupt influences of concentrated wealth."

The Farmers' Equity Union is entering a wedge that will destroy the robber profit system and the costly competitive system. Destroy the profit system and we will quit making millionaires and concentration of wealth will be a thing of the past.

Educate the people to be co-operators and the costly competitive system will be replaced by fraternal golden rule co-operation recognizing the brotherhood of man and the fatherhood of God. Our Equity Exchanges handle all farm produce without profit and will finally handle everything we buy and sell on the same principle. When strong enough we will buy direct from mills, mines and factories, saving the immense expense of competition. The factories, mills and mines spend hundreds of millions of dollars fighting one another for our patronage. We will organize our patronage and go where we please with it and save to the farmers all of the unnecessary expense of advertising and traveling salesmen.

The people are all powerful in politics and business when united. This is acknowledged on every side. The Farmers' Equity Union has discarded the word "can't" and is organizing and educating the farmers to be true golden rule co-operators. We are succeeding beyond expectation, having made a fine start in twelve states during the first four years. The movement is in its infancy, but promises a rapid growth in the future as we now have over one hundred Exchanges demonstrating the practicability of our plan and the justice of our cause. Our Union at Mott, N. D., has over two hundred members who are carrying out intelligently and successfully our plan of co-operation and they have saved nearly fifteen thousand dollars the first two years. This is being done or something similar to it, at over one hundred good markets.

When a few hundred Exchanges are co-operating fully in buying and selling, the benefits will be much greater and more apparent, and these object lessons will speak louder than volumes written on the subject. We can count on the "intelligent, God-fearing people to settle safely and sanely the trust problems" which now threaten the people with industrial slavery but we must organize and educate them more thoroughly year by year to be golden rule co-operators. Industrial Unions of the wage earners and wealth producers on the plan of the Equity Union is the only solution of the millionaire trust problem.

A NATIONAL UNION OF FARMERS

The Farmers' Equity Union must be national in scope to fulfill its true mission. The combinations with which the farmers contend are nation wide and grow stronger every year. A National Union is absolutely necessary to cope with the beef, grain, cotton and fruit and produce combination, and with the salt, sugar, match, oil, clothing, shoes and machine trusts. Under one National Head must be united cotton, stock, grain, dairy, and fruit and produce farmers. The millions of farmers must unite for self-protection and self-defense in the business world. Combinations of the people into industrial unions are necessary to cope with the combinations of the few. This is our only defense against the "Money Kings" who demand and secure more and more tribute from us as their industrial slaves. Only a National Union of farmers can throw off the yoke of bondage put upon us by the unholy combinations which fix arbritrary and extortionate prices on every thing we buy, and low, unjust prices on our crops every time we succeed, with divine blessings, in raising a bumper crop.

Every farmer must be induced to join a National Union, to quit work the first Saturday of each month and attend an Equity Union meeting and to pay the National Union one dollar per year national dues. When one million farmers are organized and educated to do these very simple, easy things faithfully, economic freedom will be in sight for our nation, and the unholy combinations by the "Money Kings" will see their doom written in letters of fire on the dome of heaven.

The Farmers' Equity Union is going about this work of organization in a sane, practical way. We begin on the foundation and we believe we are doing good, substantial work that is bound to succeed more and more as it is understood. A good strong local Union with one or two hundred members at a good shipping point is our first move on each locality. Teaching farmers the principles of true blue golden rule co-operation by lectures and literature, and a weekly co-operative paper, are of great importance. Farmers can be educated to be more fraternal and finally true co-operators. They must be educated from the capitalistic idea of big profits, big per cents and big dividends on dollars invested. Herein lies the power of the "Money Kings." We next organize a stock company, chartered by the state for co-operative buying and selling. Here golden rule co-operation is demonstrated in our plan. Stockholders are equal in vote and the majority rules. It is a peoples' organization. The people are educated to run their own business without Mr. Capitalist. Each patron puts in his part of the working capital, and only five per cent is paid for the use of the working capital while only patrons draw out the earnings of the Exchange. This is right, since patronage (not dollars) makes the earnings.

We bid for patronage and there is no grabbing for the stock. No member can haul his grain to an old line elevator, the enemy, and then come in and draw out a lot of the earnings of the Exchange made by the patronage of loyal membres. Cash is counted back to loyal members for their patronage. Cash talks! It is a convincing argument and outsiders will be reached by this argument wherever it is demonstrated. The outsider who is foolish or stubborn will finally be licked into line. We hold over him the lash of the old profit system, but we also hold open our door until he finally has sense enough to take refuge in our co-operators' camp.

The poor man as well as the rich is benefited if he is a member. He is allowed to come in on easy terms. He can take one share of twenty-five dollars and pay when he has something to sell. On our plan he finally owns four shares and then gets back the twenty-five dollars he put in, for his patronage. Our plan unites farmers. Study it thoroughly in our text book. This book will put thousands of dollars in the farmers' pockets who study its plan of co-operation and understand it thoroughly.

At some of our strongest unions we now have nearly three hundred farmers united and their financial standing in the business world will soon be thirty thousand dollars as each member finally have exactly one hundred dollars in the exchange. When five hundred or one thousand of these exchanges are linked together for co-operative buying and selling, they will have a wonderful power in the business world and show results which will astonish the natives. Then it will be easy for them unitedly to take the entire output of mills, mines and factories. Every one of them will have financial standing in the business world. They will be rated in Dunn and Bradstreet the same as a mercantile or manufacturing company.

The Equity Exchanges owned by the farmers will take the place of jobbers, wholesalers and retailers. All profit-takers will be eliminated. We will save the heavy and unnecessary expense of advertising and of travling salesmen. We will do away with the costly fight in the commercial world for our patronage. The war in the competitive world will be ended. When the patronage of one thousand Equity Exchanges are thoroughly organized, we will go where we please with it, and the power of the trusts to rob the farmers will be broken. Combination, concentration and co-operation are the three C's which will bring economic freedom to millions of farmers.

We will center the wheat of our big elevators with individual mills of our own, so they can run steadily one hundred forty-four hours each week and fifty-two weeks each year. We will brand "Equity Union" on every sack of flour and thus guarantee the sale of all of our flour made by our mills to union farmers and to union labor people. Our market must be organized to receive the output of our mills, so that the flour trust cannot bribe nor buy the people, our market. Wealth makers must unite in industrial unions against wealth takers.

The people are all powerful when united. Fraternalism, the twin sister of co-operation, is all important in this movement. These two must go hand in hand. Teach the people to be fraternal co-operators. This is our only hope and can and must be done. Every intelligent farmer must be educated and induced to help. Farmers, let us no longer be pessimistic standpatters, but optimistic co-operators.

We will win if we push forward, because we are right and deserve to win. A National Union of farmers must be grown. We are making the foundation strong and substantial. It is founded on the principle of true golden rule co-operation.

As our Equity Exchanges demonstrate this principle, they will be more and more in favor with the people. We can work out a business proposition worth hundreds of millions of dollars to producers and consumers of food products. We can buy all farm machinery at one-half of the present prices. We can work out a system of distribution, that will guarantee equitable prices to both producers and consumers.

We have just read in a paper where one man paid ten cents for a single apple, and in the same paper where thousands of bushels of good apples were rotting in orchards in Wisconsin. If a million farmers were united and reading the Equity Union Exchange weekly, how easily we could locate the supply and also the place of demand for apples, peaches, cabbage, potatoes, onions, etc., and distribute them to the advantage of all of our members.

Our Exchanges are beginning to distribute their products directly to each other. We find at this writing that peaches are fifty cents per bushel in one locality and that the price is prohibitive for common people in another place. The fear of grafters and politicans is keeping some farmers out of our Union. But this need not deter any farmer from joining. We are an industrial Union and will never attempt to become a political party. We are determined to keep out of party politics. No set of politicians nor political party will ever be allowed to dominate our unions. We are also fixed to look after the Mr. Grafters. Our by-laws require that every person who handles money shall be put under a bond by a reliable bonding company, and an expert bookkeeper examines the books every six months and reports to the stockholders of our Exchanges, so that our business is run as safely as any business can be. A good bonding company stands between the people and all danger. A cut-off is required at each elevator in January and June of each year, and all books are examined by an expert bookkeeper and a report made to the stockholders.

The strong point in favor of each of our elevators, is the fact that it has a National Union behind it. This National Union educates each member to stick to his own exchange. It goes after the trade of every farmer. The grain elevator men have tried repeatedly to buy our members and failed. At Liberal, Kansas, for one week they paid three cents above the market in July 1912, but our members hauled their grain to the Liberal Equity Exchange in spite of the bribe offered by the old line companies. One member hauled his grain amounting to 7,000 bushels of good wheat to our elevator when the enemy was paying three cents a bushel above the market. In August 1914, at Bird City, Kansas, the enemy put the price seven cents above the market and failed to get a single Equity Union member to haul them their grain. This all shows the grand work of eduiation which the National Union has given its members. We continually stir, arouse and encourage every farmer to come in and be a golden rule co-operator. We do not dictate as a National Union but we look after the management of every one of our Exchanges. We look for good, honest, capable men to manage the Exchanges, and we will finally have the national government to establish a school, where our brightest country boys can be educated and trained to run our Exchanges in a business way.

We are endeavoring to have every manager bonded by a reliable bonding company. We are also working for a uniform set of books in all of our Exchanges, so that an expert can examine them easily and thoroughly twice each year, and oftener if necessary. The National Union is also working over time, to concentrate the business of all these Exchanges with mills, mines and factories to the advantage of each and every exchange, and every individual member.

Farmers, why fight our battles single-handed any longer, with great combinations, when we can unite and have the power of one million men. Start in now at your town and build steadily with the rest of us and in seven years we will have the million united and reap the benefits of our strenuous efforts and leave our children free from tyrannical "Money Kings," who seek to keep them forever in industrial slavery. I want no better momument to my memory, than a great Industrial Union of farmers, to which I have given my life and did my best to start and build up on true blue golden rule principles. **To this end is my life devoted.**

MONOPLY BY THE PEOPLE

In the Saturday Evening Post of August 26, 1911, Gifford Pinchot said, "If conditions are such in Alaska that development without monoply is impossible, then let us, the people of the United States, become our own monoplists and hold the monoplies in our hands."

This doctrine is growing more popular every day, and will finally apply throughout our country. The people are awakening slowly to the fact that they could live and be happy and prosperous, without a single capitalists if the people were more intelligent, moral and fraternal, and true golden rule co-operators. The doctrine is being instilled into the minds and hearts of the people from many sources that they can and ought to run the politics and business of this country without "Bosses" or "Millionaire Magnates."

There is a real cause for this awakening of the people. We seem to allow evils in our country to become full grown before we tackle them. The people are viewing, however, with great alarm in some sections, the fact that a few selfish "Money Kings" are gaining more and more power every year over the distribution of their food products, clothing and shelter, the necessities of life. No monarch ever had more absolute power to compel his subjects to pay tribute than have the "Money Kings" of so called free America.

They combine and monopolize our sugar, salt, matches, oil, meat, flour—all food products—and our clothing and lumber. They act as toll-gates and demand tribute from those coming and going. They price down on producers and up on consumers the necessaries of life. The raw material produced by farmers is priced as low as possible, and the finished product is raised to the highest notch on consumers. These facts are more evident every day. It means tyrannical slavery to millions of honest industrious farmers and their entire families. The farmers as a class are the great producers of our immense wealth and also, being great consumers, they suffer more from these economic conditions than any other class.

The "Money Kings" are the most heartless, cold-blooded and tyrannical of monarchs. The love of money which is the root of all evil, seems to crush out of their bosom all love for humanity. Gold is the meanest god ever worshipped by man and we have more of this worship in Christian America than in any heathen land. These conditions are compelling every lover of humanity to look for an avenue of escape for the people. Many farmers are aware of, "a screw loose somewhere," but are unable to locate it. Others understand the situation but see no remedy.

We believe Gifford Pinchot's "Monoply by the People" must not be confined to Alaska, but must be applied to the whole United States. We believe in co-operative ownership or "monoply by the people." Mines, factories and rail roads should be owned co-operatively by the people who operate them. The mine-workers should own and operate the mines and elect their own boses. They should at least own sixty per cent of the capital while the government owns the other forty per cent, and never allow capital invested in the mines to draw over two per cent stock dividends. The factory workers must become fraternal co-operators and own at least sixty per cent of the capital invested in factories, machinery and tools with which they work, while Uncle Sam owns the other forty per cent. The rail roads should be owned by the millions of men and women who operate them in unison with the national government. This movement is already on foot and laboring men are being persuaded to take stock in mines, factories and rail roads. When they own co-operatively in conjunction with the government, the factories, mines and rail roads, there will be no more tyrannical capitalists, and no more strife between labor and capitalists. We will have no more monopoly by the few, but "Monoply by the People."

This desirable state will be brought about by industrial unions, which teach the Fatherhood of God and the Brotherhood of man. **Economic freedom is only possible to an intelligent, moral, fraternal people.** But God loves the common people, and has set in motion in our beloved land mighty forces for their freedom. The Truth shall make them free and they shall be free indeed.

The principles of true golden rule co-operation must be instilled into the hearts and minds of a strong majority of the people. "The farmers of America will be owned and controlled by the trusts within the next thirty years if they do not co-operate," was the warning issued by President T. J. Brooks of Tennessee, in his address at the opening of the National Farmers' Union at Shawnee, Oklahoma, September 4th, 1911.

United States Senator T. P. Gore, in the same meeting, said, "You farmers will be buried deeper in misery than ever if you do not co-operate."

As the miners, factory-workers and railroad operators must monopolize the mines, factories and railroads, so farmers must monopolize the channel or means of distribution of their valuable products direct to the consumers.

The ware houses, coal sheds, elevators and all necessary equipments for handling our produce must be owned co-operatively by farmers in stead of by trusts. The farmers must own the elevators and ware houses at every good shipping point in the country and finally own distributing plants in the central market for the distribution of their products. Our fat stock must go to our own stock yards, where they are comfortably housed and fed at reasonable rates, and a steady supply furnished to the central markets at a steady equitable price. Our grain must be shipped to our own terminal elevators and mills and only put on the central markets as the demands calls for it at a just price to producers and a reasonable price to consumers. We must have our own milk station and co-operative creameries in the country or at central points and wholesale plants in the cities for the honest, intelligent, economic distribution of dairy products.

Fruit and produce of all kinds must be handled in the same co-operative way. Farmers must monopolize each channel now owned and controlled by trusts, gamblers, grafters, profit-takers and price-hammerers. A scientific, economical plan of distribution is being worked out by our Union which will be worth hundreds of millions of dollars to both producers and consumers.

Farmers, we ought to have done this thirty years ago. Do not neglect nor refuse to act any longer. We have piled our hard earned cash into the hands of the monopolies long enough. We, the people, will become the monopolies and hold them in our hands. The Equity Union is starting "a peoples' monoply," as the only remedy for the present monopolies controlled by the "Money Kings."

A thorough organization of the farmers will break the power of every trust to rob the people. The big concerns are not much alarmed about us yet. They bank on the old claim that the farmers cannot be organized. They say, "They won't stick." The Equity Union banks on the idea that farmers are human beings with minds and souls capable of education. We have a practical plan of co-operation which, when carried out, counts cash right out under each man's nose as a result of his co-operation. Only small dividends from three to five per cent are declared on the stock subscribed, which enables each exchange to pay back for patronage some amount almost every year. Mr. Capitalist is kept out, and only patrons can draw out the earnings of each company. The fundamental idea of the **Equity Union is, that the earnings of the stock company belong to the patrons.**

Patronage makes the earnings. Stockholders cannot draw their grain to an old line elevator and then draw out the earnings of their own elevator made by loyal members who furnished the patronage. We believe this provision is wise and just, and we know that it is holding our members together, and believe also that it will finally make the monopoly we seek to make by the people. Not only will the people be their own monopoly in Alaska, but throughout the United States, through their industrial unions.

Westward the Star of Empire has taken its course, for the last century, but now backward from the golden shore of the Pacific comes the wave of reform overthrowing political bosses and checking the power of Millionaire Magnates, and teaching the people how to curb, regulate and control or eliminate the monopolies. These reforms are advancing rapidly. We may not be drifting to political socialism, but to "Monopoly by the People," through industrial unions which promote the intelligence, morality and frateranlism of their members and make them golden rule co-operators.

THE BIGOTS AND REFORMERS

United States Senator John Sharp Williams said in a speech in the senate: "Why, my friends, men in religion, men in politics, men in trades, have been afraid of freedom ever since the world began. God Almighty seems to be the only being anywhere, who is not afraid of freedom, and not afraid to give it to his creatures. He gives it to such an extent that he lets us go wrong if we will. From the beginning religious bigots have been afraid of it, political bigots have been afraid of it and industrial bigots have been afraid of it, and yet, whenever it comes, we find it stimulates human enterprise, human intelligence, human ambition and human industry to such an extent that it more than compensates for what seems to be the plain palpable and obvious immediate losses."

In our opinion, this is one of the most sensible things ever said in the United States Senate. Bigots have always opposed freedom for the people. True, honest reformers have always found the bigots their bitterest enemies. The humble Nazarene found them strongly entrenched behind the walls of ignorance, prejudice, Phariseeism and inherited tendencies.

Religious and political freedom have been won by reformers, who were despised and spit upon by bigots. But the history of the world shows that bigots have gone down in disgraceful defeat, while the humble, despised reformers eventually have monuments erected to their honor and memory and are enshrined in the hearts and memories of the common people with love and adoration.

The great battle now on in our country for economic freedom of the common people is no exception to the general rule. Bigots, selfish interests, led by political bosses are all arranged in solid phalanx against every movement for true co-operation that will bring economic freedom to the people. Victory can come only as the people are educated to be true blue co-operators. The religious and political bigots were overthrown by the schools and churches, which led to intelligence, morality and fraternalism among the people. The same remedy will bring economic freedom and overthrow every industrial bigot.

The Farmers' Equity Union is a reform organization now being firmly planted in twelve states. It has a definite, practical plan of true blue golden rule co-operation, that will finally overthrow every industrial bigot, who now so strenuously opposes the movement. It depends largely for success on a continual campaign of organization and education. The National Union has no capital whatever. It has an entrance fee of two dollars and one dollar for the paper and one dollar annual dues for its support. This money is used for equipping and running a national headquarters, printing a weekly paper, literature, circular letters by which we seek to keep in touch with every member and which we use to put more and more honest, able lecturers in the field to hold meetings and arouse, encourage and instruct the people in the principles of our up-to-date organization.

Our work has resulted in co-operative business unions at over one hundred good markets which are carrying out our principles of co-operation. We have been working strenuously winter and summer, early and late, at these one hundred markets for nearly four years, laboring incessantly to shut the mouths of bigots, standpatters and pessimists. This has been stone-masonry work, but the foundation had to be laid. Individual members have been reached with a weekly co-operative paper, literature and lectures for four years. Our co-operative paper has educated whole families on our principles. The first Saturday of each month is Equity Union day and is beginning to be religiously observed by members and their families. The one dollar annual dues paid into our treasury is the sinews of war which we are using to drive back all oppositions to our movement.

This work cannot be done without money. If it costs nothing, it will be so poor that it will be worth nothing. Success in this battle is only possible as we carry on a continual campaign of organization and education among the people who ought to unite. The more money we have the more rapid progress we make in building up our Union. There must be a continual onslaught against the bigots and the supremely selfish millionaires, who now hold the millions of farmers down in industrial slavery. The farmers must be aroused, encouraged and instructed by the very best speakers to strike the blow for economic freedom. This will take money and without money the movement will be a failure.

Our new co-operative elevators now being built by our members, will not cost them a dollar if they adhere to Equity Union principles, as we believe they will. Each member pays in from twenty-five to one hundred dollars as his part of the working capital. No member can hold over four shares and no member can draw out anything till he has four shares, the limit. Hence each member is equal in capital and vote. We buy on a safe margin and never declare over five per cent dividends on the stock subscribed. This always gives from three to twelve per cent patronage dividend to prorate back to each stockholder for his patronage.

This encourages patronage and pays for the elevator out of the farmers' patronage, which they give the middle men. The money each member puts in the Union and in the elevator is finally counted back to him in cash where he can see it. The elevator and union do not cost him one cent, but come back to him in cash which he now loses by listening to industrial bigots, who are opposed to co-operation and all reformed movements which bring freedom to the people. Outsiders finally see the dollars counted out to members and come in and stay with the union. Stock is always for sale when outsiders wish to come in. We want ninety per cent of the farmers united if possible, and we want the poor man to have the full benefit of co-operation as well as the rich man. He can take one share and pay when he sells the crop. Only a few years of patronage will give him three more shares. And another year will give him back the twenty-five dollars he put in. Then he has four shares which cost him nothing. Co-operation has conserved his patronage and instead of giving the millionaire twenty-five to one hundred dollars, co-operation counts it back into his pocket.

The co-operation of hundreds of farmers at each good town is worth thousands of dollars to each locality annually. The co-operation of thousands of Equity Exchanges will be worth hundreds of millions of dollars each year. Our products will all go to the central markets through channels owned and controlled by farmers instead of speculators, profit-takers and grafters. These crops will be handled at actual cost without profit, direct from producer to consumer. The surplus will be held in the country in the safest, and cheapest places of storage. Supply will be regulated to demand and fair, steady prices maintained on all farm produce.

Thousands of exchanges must buy farm machinery co-operatively in large job lots direct from factories, and reduce the price fifty per cent. A fair, equitable price on all we buy and sell is in easy reach of every farmer through our union, if he will become a member. These are only a few out of many benefits which the Equity Union will bring every farmer who joins. Economic freedom will be assured to all farmers, when one million do four easy things, viz., join the Equity Union, read our paper weekly, meet regularly the first Saturday of every month and put one dollar a year into our national treasury on November the first. The country will then bloom as the Rose of Sharon. Conversation of patronage will enable us to conserve moisture and fertility by giving us financial means for such purposes and making our crops sure. Sure equitable prices will bring sure crops to some extent, and make the farmers' business the best business in our country, as it ought to be.

Brother members, write to your agricultural paper, tell them what the Equity Union is doing in your community and ask them to advocate co-operative buying and selling on the Equity Union plan. If the editor is one of the bigots, tell him you prefer a paper which teaches the farmer to sell his crops co-operatively; and that the present teachings of too many agricultural papers only helps railroads, handlers, speculators and others to make millions of dollars out of our best crops, while farmers sell them below actual cost of production. The Equity Union Exchange is a true blue advocate for golden rule co-operation. It is doing more for co-operation than all of the other agricultural papers combined.

THE COST OF DISTRIBUTION

B. F. Yoakum, chairman of the Executive Committee of the Frisco Lines, says the big problem before the people is the high cost of selling. He points out that the selling cost of farm products frequently amounts to one hundred and four per cent. This is the problem which the Farmers' Equity Union is endeavoring to solve as we grow stronger year by year. We are powerless to do anything whatever with these great problems while more than three fourths of the farmrs are on the outside watching for results by the one-fourth. The farmer, on the outside, who points the finger of scorn and declares we are accomplishing nothing, is himself largely responsible for our weakness, from the fact that his help and influence to our cause has been refused and his influence against us, while we have battled manfully for years to improve conditions for him and ourselves.

The first great work of the Equity Union is to make five hundred strong links at five hundred good shipping points. We are making them so strong that they are as enduring as Gibralter. The bulwark of Equity Union is strong local unions at our best shipping points, backed and supported by two hundred or more intelligent farmers and their families, who are thoroughly instilled in mind and heart with the principles of golden rule co-operation which leads them to have more love for humanity than for the almighty dollar. Show me such a community and I will in a few years show you a locality with good roads, schools, churches and beautiful homes. You will see a prosperous, progressive people, with the average per cent of peace and happiness very high.

We must cease the everlasting grind for the almighty dollar, and center our efforts for the uplift of humanity, and yet the question of intelligent distribution and co-operative buying and selling, enter so largely into the very woof and web of humanity's needs, wellfare and happiness, that these questions continually press themselves to the front. We must interest the people in the discussion of material things, but not leave them there. The right use of the dollar is more important than making it. Greed and selfishness, the great bane of society and of the business and political worlds must be supplanted by fraternalism and love for humanity.

Mr. Yoakum's question is being discussed in all our literature and in our Equity Union meetings every month. We are sure to work out a more intelligent system of distribution of our valuable products to the advantage of both producer and consumer.

A Minnesota farmer a few years ago, complained that he spent the price of five bushels of potatoes to get one basket of Missouri peaches and at the same time the Missourians were giving five baskets of peaches for one bushel of Minnesota potatoes. Potatoes often leave the producers hands at twenty cents per bushel and retail at twenty cents per peck in many of our central markets. Cabbage often brings the Wisconsin farmer only five dollars per ton while retailing to consumers in our central markets at four cents per pound or eighty dollars per ton.

Apples often sell from the farms at twenty cents per bushel, and retail to consumers in our central market at two dollars per bushel. The producers receive less than three cents per quart for milk, in many sections, while consumers pay from eight to ten cents per quart in the cities. Seventy-five cent wheat from the farmers' hands often costs the consumers two dollars per bushel when bought at retail bakeries in our cities after the speculators get the bulk of the new crop in their hands and boost the price on a long suffering public.

One year, I saw a fine bunch of fat cattle in the spring, just finished off in the spring for the market. They were fancy stock. The farmer had put his entire corn crop into them and bought some besides. His price was so low when he sold the cattle that he actually lost money on every one of them. If he had received fifty per cent of the consumers price, he would have made a fair profit. It costs less than fifty dollars to make an eight foot self-binder, and our members in the Northwest are paying from $150 up to $170 for that same self-binder. They pay fifty dollars for material and labor and $130 for transportation and distribution. The material and labor for best wagons cost less than $40 and retail for $80 and so on with all farm machinery. These are problems not only for discussion, but for solution. They can be solved by organization, education and co-operation.

When organized more thoroughly, we want hundreds of lecturers every month, to visit our large Equity Unions in the twelve states and lecture on these and kindred subjects. When thoroughly organized at five hundred good towns, how easily and economically we can reach all of the local unions, once every month with the very best and most expert lecturers on all topics of special importance and interest to us as farmers. We want the members of our union to read the Equity Union Exchange regularly and religiously every week, so there will be more and more uniformity of action in buying and selling, and more and more co-operation. The power of each individual farmer will be increased by one million farmers in a union. This point must be continually hammered so that every farmer will be impressed with the power he has as an individual member of a powerful Union.

The boy in the coal mine is mistreated by the boss. Can he have it made right? Yes. He can come right up out of that mine and lay his grievance before the grievance committee of the miners' union, and make Mr. Boss step right up on the carpet and give an account of his treatment of that boy. How can he do it? How does a mere boy driving a mule down in a coal mine happen to have so much power to get his rights? I will tell you. He is a member of the United Mine Workers of America, and there are five hundred thousand of them in our country, and every one of them will come out of the mine if necessary, and stand by that boy for his rights. It is the power which he has as an individual member of a powerful union, which gives him protection.

Brother farmer, listen! You are an American citizen! You ought to be glad and proud of the fact. But suppose you travel over in Russia. If you do, the power and influence of that great government will be brought to bear to protect you from all harm, all injury and all insult. Why will they do it? Because they know that over on this side of the Atlantic Ocean there is a big union of one hundred million people who will stand with you for your rights as an American citizen. It is the power which you have as an individual member of a powerful union which gives you protection.

Brother farmer, will you not lay aside every prejudice and suspicion, indifference neglect and carlessness, and especially that horrible word "can't," and begin now to help me build a great industrial Union of farmers and their families, which will give each member the power of millions, and that will take up and solve intelligently and successfully the great economic question now before the American farmer?

Mr. Yoakum is right. The high cost of selling from the farmers to the consumers, and from the factories to the farmers, are great economic questions which can be readily and surely solved on the plan and principles of the Farmers' Equity Union.

THE EQUITY A SUCCESS

We are organizing the farmers in the Equity Union as rapidly as we can get a hearing, from the fact that it is easily shown that the organization is self supporting. The organization and the elevator purchased, do not cost the members one cent, when they get together and carry out fully our plan of co-operation. If the member takes one share of twenty-five dollars in the elevator and patronizes it, on the Equity Union plan of prorating back all profits, there will soon be twenty-five dollars coming to him for his patronage. A second share is then given him and finally a third and fourth share. He now holds the limit—four shares. When there is another twenty-five dollars coming to him for patronage, as there will be, the original twenty-five dollars is paid back to him in cash and he holds four shares, and neither the union nor his four shares have cost him one cent. Co-operation has conserved his patronage and given him four shares in the exchange. He has received as much, or a little more, for his grain than he would have received, if no farmers elevator had been at his station. His weights, grading and dockage have been just and right, and he owns four shares worth one hundred dollars in cash. After this all profits are paid back to him in cash year after year. All he buys and sells are really handled at cost, and he has the great satisfaction of knowing that he and his brother farmers are their own capitalists, and privileged to run their own business without dictation from the millionaires.

Our pass word is "Mind your own business." The Equity Union is educating the farmers to run their own business co-operatively and safely. Our members are learning to trust themselves and to trust one another. They are learning to attend to their own business without piling up the millions of dollars they earn each year in the hands of the capitalists. They need plenty of capital, and since they are learning to conserve their patronage they have plenty of it. In the Equity Union they are being fully convinced that they can get along very well without Mr. Capitalist, Mr. Grafter, Mr. Extortioner and Mr. Profit-Taker.

The directors audit the books regularly and carefully and have a cut-off in January and June of each year, when an expert bookkeeper inspects the books and reports to the stockholders. The directors and stockholders look after the business keenly and carefully all the time, so that they know how it is running. Many of our members in 1912 and 1913 had the $100 paid back to them each year. I am astonished and chagrined, when I think how many old line elevators the farmers have paid for, over and over again in the West and Northwest by their patronage, and the millionaires still own them and continue to make the farmers pay for them, every time they have a bumper crop.

The worst thing I find, in the West, is the so-called farmers' elevators on the old-line-capitalistic-skin-game-plan. A few farmers build an elevator and in a good year they declare fifty and sixty per cent dividends on the stock subscribed and **no shares for sale**. In many cases they are wolves in sheep's clothing. Some of them are Sunday school superintendents! It is enough to make angels weep to see the greed and selfishness of some men. Fifty per cent dividends by the beef trust, grain trust, milk trust or oil trust are no worse in principle than the same dividends by a farmers' elevator company, and no stock for sale. Every farmer can take at least one share in an Equity Exchange, however poor he may be, and share the benefits of co-operation and finally have the same voice in the management in the business and the original twenty-five dollars will be paid back to him so that the elevator does not cost him one cent. His one dollar national dues are also paid out of the profits made on his patronage for him, so that neither the organization nor elevator costs the member one cent. It is all self supporting when the Equity Union co-operative plan is carried out by the members, and the poor farmer has benefits as well as the rich man.

If the editors of our agricultural papers would study true blue golden rule co-operation and advocate it in every issue as some of them are doing, and educate the millions of farmers along this line, economic freedom would come to the people in one decade. They would learn how and be encouraged to strike the blow that would break the fetters that now bind them to the robber profit system, and be free forever from the capitalistic, tyranny and industrial slavery.

Every Equity Union member ought to demand this service from his agricultural paper, or stop supporting it. Let us have farm papers that educate we farmers to conserve our patronage as well as moisture and fertility. They are all equal in importance. Farm papers which are controlled by capitalists and combinations should have no place in a farmer's home. They are his enemies and should not have his support. The agricultural papers have it in their power to make every farmer a co-operator and in this enlightened day are responsible for farmers' ignorance on the subject.

Farmers, listen! Co-operation is the loaded gun which lies by your side, with which to defend your self against every foe. The equipments for using it will not cost you one cent. Will you not awaken from your sleep and use this weapon of defense? Let us get together a million strong and fire the gun for freedom that shall be heard around the world.

THE FATHERHOOD OF GOD AND BROTHERHOOD OF MAN

The Farmers' Equity Union is founded on a belief in the Fatherhood of God and Brotherhood of Man. Love for humanity is a fundamental principle of our union. We are working for golden rule co-operation. Fraternalism among the millions of farmers is our chief object. "If we love not our brother whom we have seen, how can we say we love God, Whom we have not seen?" There is no danger of loving our fellowmen too much. The Union that teaches unselfishness fraternalism among men is a benefactor of the race. Abraham Lincoln said: "God must love the common people are he would not have made so many of them."

The Equity Union is for the common people, first, last and all the time. You will find a larger per cent of this class among the farmers than you will in any other class. As a class, the farmers are honest, industrious and economical. They produce more than one-half of all the immense wealth of our rich country, every year, but they do not get one third of it. They are the great wealth making class, but they have to protect themselves from the hordes of wealth takers. Hence they are exploited on every side by grafters, speculators, extortioners and combinations.

The condition of crops raised by these industrious toilers is, throughout the season, a prolific topic of conservation and of discussion in all of our news papers. Very little is said about the crops or farm machinery or dry goods or groceries. It is all about the success of the farmer or his failure. I wonder why? The great question seems to be in the business world among the wealth takers—how can we educate these industrial slaves to produce bumper crops, out of which railroads, millers, packers, speculators, handlers and combinations can reap a rich harvest of millions of dollars?

But, brother farmers, let us not lose sight of the fact that inactivity, neglect, indifference and want of co-operation among ourselves is responsible for every bad condition in our business. Mr. Wealth-taker will have no power or special privilege when farmers unite in the Equtiy Union and carry out fully its practical plan of co-operation. Mr. Exploiter, Mr. Grafter, Mr. Profit-Taker and Mr. Speculator will all be out of a job when a million farmers are golden rule co-operators. A million organized farmers will produce bumper crops for the benefit of producers and consumers instead of for the sole benefit of a few millionaires.

We are perfectly willing that railroads, millers and handlers, and every one engaged in a legitimate business shall be well paid for good, honest service, but we insist that the separation of farmers, their indifference to the power and benefits of organization and want of fraternal co-operation, is their great weakness and makes them an easy mark for every combination. As long as farmers neglect or refuse to organize, the present conditions will prevail.

The Equity Union is teaching the Fatherhood of God and Brotherhood of Man. The general uplift of humanity is our primary object. We must battle down the walls of selfishness, greed, ignorance, suspicion and prejudice in farmers themselves by a continual campaign of education. Hundreds of clear(forceful, honest, sincere speakers must be kept in the field agitating and educating. Literature in both English and German must be circulated by the ton. A good co-operative paper teaching farmers to be brothers must go to every members' home. We will continue to insist that this grand movement can only succeed as farmers are educated to be golden rule co-operators. It is impossible to do this without money. A regular revenue must be provided for educational purposes. Every Equity Exchange must pay the one dollar national dues on November 1st, of each year for every stockholder and charge the same to his account. As the parent is worthy of support by every child, so our National Union deserves support from every local and district union. One dollar per year from each member, is only a trifle and will work no hardship upon anyone, but when enough members pay in the one dollar regularly each year, a good live headquarters can be equipped a good weekly co-operative papers and plenty of literature printed, and more and more field workers can be employed and kept busy educating the farmers up to the high plane of true blue golden rule co-operation.

As a great manufacturing business depends largely for success, upon its army of intelligent, hustling, traveling salesmen, so the Equity Union must look to its energetic, wide awake, whole-souled canvassers, organizers and lecturers. We must recognize and appreciate the services and self-sacrifice of our able field workers. Their work is fundamental. Without them our movement will fail. Theirs is a continual battle against ignorance, preconceived ideas, suspicion, prejudice and inherited tendencies. They must make constant courageous warfare against strongly intrenched systems of graft, extortion and robbery of the people. The field workers are indispensable, and those who stay with it and succeed deserve all encouragement and sympathy, we can possibly give them, and they must be guaranteed a fair salary and all necessary expenses.

One great advantage the Equity Union has over nearly every other farm organization is the fact that it has a definite, practical business plan of co-operation to present to the farmers, which we know is being carried out at one hundred good markets to the advantage of every good member and which we know is uniting the farmers and keeping them united. Our paper and the Equity Text Book, all of our literature, our speakers and organizers are centering their efforts in educating and persuading the farmers to organize and carry out this practical business plan of co-operation.

Our plan begins by showing the members dollars for their patronage at our Equity Exchanges. We buy and sell on a safe margin and never pay over five per cent on the stock subscribed. This enables us to pay every patron who is a stockholder some cash for his partonage once each year. This is the profit of co-operation. Most other profits are the robbery of the wealth maker by the wealth taker. Cash for patronage is simply putting back in the pocket of the wealth producer what rightly and justly belongs to him, instead of allowing the profit-taker to rob him. Cash paid for patronage is a practical demonstration to each farmer that it pays to co-operate. It is cash that talks.

It is a little stronger argument than any lecturer or writer could give in mere words or theory. The dullest man or woman farmer understands it. They can count it in cash. This is the only way to unite the farmers in a National Union. Thorough education on these principles of true co-operation and a practical demonstration by actual cash counted by the farmers themselves will eventually unite ninety-five per cent of the best farmers at each good shipping station and keep them united. Buy on a safe margin, have good honest management and prorate back in cash all earnings over five per cent on the stock subscribed, and you will shut out the capitalist, and the profit-taker; and farmers, the wealth-makers will get their own.

There is no end to the possibilities of this movement. Five hundred links in our Equity Exchange will show such power and benefit that a million farmers will unite in a few years. There is, however, a great difference in the farmers of different localities. In some places they rush right into the Union and are ready and eager to learn its plans and principles, and they are courageous enough to unite and carry them out successfully. This is especially true in the great West. It is western style, and speaks well for their energy, grit and gumption. The West seems to develop progressive men and women, and will finally rule America in politics and business. The center of power will no longer be the little Northeast, but the great West.

At Bowman, N. D., in July 1914, we organized a Union with 189 members. A majority of the members paid one share of $25 when they threshed their grain. They bought an elevator with flour house and coal sheds, hired a good manager and centered their trade together. They buy and sell on the same margins as their competitors, unless their competitors pay more than the market price will justify. They will prorate all net earnings to stockholders according to patronage, not in cash, but in shares. They will pay for their elevator in the handling of one good crop. When North Dakota farmers are shown a good proposition they do not wait and see, but take hold of it in earnest and carry out our plan. There is not a profit-taker nor standpatter among them.

As the Equity Union grows the profit-taker will be relegated to the rear and the wealth producer will come into his own. Millionaires and industrial slaves will be less numerous. The common people (God's kings and queens) will be more intelligent, moral and fraternal, more firmly united by the bonds of love, friendship and kindness, the grand principle of the Fatherhood of God and the Brotherhood of Man will be firmly established in the hearts and minds of the people.

RELATIVE VALUES AND PRICE-FIXING

One of the great economic questions now up for solution by the American people is the relative value of all commodities produced by our real wealth producers. Gold is considered as the standard of all values, but we contend that gold has no independent value. Nothing of real value can be produced except by labor combined with skill, hence the processes of commerce in effect consist of trading the labor and skill contained in one commodity for the labor and skill contained in another commodity. In the suspended exchange, however, it is necessary to have a medium of Exchange. The real standard of value, and therefore the medium of exchange, is a day's work combined with skill, but as labor and skill cannot be measured or weighed and done up in a package, it is necessary to use as a medium of exchange some suitable commodity, like gold which cannot be produced except by labor combined with skill. Therefore the amount of gold which can be produced by a day's work combined with skill is convenient to represent the unit of value.

As inventions and discoveries enables this quanity of gold produced with less labor and skill, the comparative values of gold with other commodities, which are produced also by labor and skill, will decrease unless invention and discoveries enable a day's work combined with skill to produce all the other commodities at a light reduction in cost, which has been the case.

So, if values were left to be settled by the natural law, there would be little if any, apparent reduction in the value of gold. But the difficulty today is just here—there has grown up in our country gigantic industrial monopolies, which are able to set at defiance the natural law of values and place such values as they see fit on their products. These monopolies having placed artificial values upon their products, a fair exchange cannot be made unless an equally high value is placed upon labor and all other products, which are produced outisde the monopolies.

It is apparent then to all who think and view the situation intelligently, that the only hope of the one hundred million wealth makers of our country is to organize into industrial unions in order to defend themselves against the one million wealth takers who dominate our commercial life in such a way as to destroy and defy the natural law of values. These industrial unions must not be dominated by any one political party, but must themselves through the power of their national union dominate all political parties as they will eventually do. The monopolies will then be relegated to the rear and the people will come to the front and rule as they have the divine right to do both in the political and business world.

These industrial unions must make much of the educational feature. Money must be provided and spent freely for this purpose. A weekly co-operative paper must reach each family, teeming with the discussion of these great economic questions—the distribution of farm products, the pricing of labor, the just distribution of billions of wealth produced each year by the people. Literature must be issued by the ton. Meetings must be held continually and hundreds of honest, able lecturers must be kept in the field constantly.

The industrial union must make strenuous efforts to promote the intelligence, morality and fraternalism of its members and educate them to be golden rule co-operators, keeping in mind the fact that economic freedom is only possible to an intelligent, moral, fraternal people. The Farmers' Equity Union of Greenville, Ill., is just such an economic industrial organization, seeking to unite the great class of people, who now work more hours, perform more labor, produce more wealth and for less money, than any other class of wealth producers.

The trusts fatten off the millions of toiling farmers and their families. A large per cent of farm products are produced by wife-labor and child-labor. Five millions of farmers are working for their board and clothes and giving all of the billions of dollars they produce annually, to the beef trust, grain trust, milk trust, fruit and produce trust, the flour, sugar, salt, match, oil, cloth, boot and shoe, lumber and other trusts. But there is hope for every farmer. The dawn of day looms above the horizon. We all acknowledge that the millions of farmers are powerful when united. We feed and clothe the world; we furnish the world the necessities of life. We have in our hands each year what the people must have or perish. We are under no moral obligation to bread the world at a price which is below the cost of production as we have done in many years in the past.

The farmers are becoming more intelligent, moral and fraternal every year. They are reading and thinking as never before. The Equity Union is calling them together in the school houses and towns, in contests, and corn shows and farmers' institutes every month. We are taking up the discussion of the great question of distribution as well as production. We appeal to the pocket book interest of the farmers which touches them in a tender place. We introduce co-operative buying and selling at each good shipping point which conserves his patronage, which he now wastes to the extent of millions of dollars.

Hundreds of Equity Exchanges must be organized. Thousands of farmers' elevator companies must come under the Equity Union banner. We must have expert salesmen to sell our grain co-operatively, and finally we will have Equity Union mills owned by our members, which will grind our own grain.

RUINOUS COMPETITION AMONG FARMERS

In March, 1904, the German steel industry consolidated into the "Steel Works Union." This combination has a yearly output of eight million tons, while the output of the United States steel corporation is ten million tons. Concerning this German steel trust, Professor Riesser, an eminent German economist, has written: "The necessity for the formation of trusts in Germany was clearly recognized in the seventies. By this means over-production and ruinously low prices were terminated."

Every trust in our country is organized for protection against "ruinously low prices," produced by either foreign or home competition. The business world has been compelled to learn the lesson that "ruinously low prices" are caused by unhealthy cut-throat competition. They have long ago substituted co-operation for competition so as to prevent "ruinously low prices." The business world combines and co-operates more and more every year. More and more of the big banks merge, more and more of the mills, mines, factories and railroads combine.

Combination and co-operation are the order of the day in the business world. Large corporations grow larger and stronger every year and are here to stay. Big business will grow bigger, not smaller. Not only manufacturers, railroads, banks and mines are uniting, but wholesalers and retailers are organizing more thoroughly every year. The time has come when the seven million farmers must meet regularly every month and discuss, consider and find out their relations to present conditions in the financial world. The seven million wage-earners are organizing all over our country to prevent "ruinously low prices."

The farmers are paying a trust price and union labor price for all they buy; labor on the farm is our most preplexing problem. To say we are opposed to all trusts and unions does not alter conditions nor solve the problem. The logical view is that big corporations are here to stay and grow larger, not smaller, and that more of the wage earners will unite to prevent "ruinously low prices." All intelligent people now recognize these facts. Then what is the logical conclusion? **Farmers must unite and co-operate** to prevent over production of any one crop and "ruinously low prices" on what they sell and high prices on what they buy. They must work out an intelligent, economical system of marketing by which the surplus is held in the country in their hands and central markets fully satisfied each day and none glutted.

Exchanges must be financed and equipped so that shipments of all products can be made as directly as possible from producers to consumers. Ruinous competition among farmers in selling their valuable crops brings "ruinously low prices," and causes us to sell below cost of production, our largest crops of finest quality. While hogs are dying with cholera all over the country, corn high priced and an actual shortage of hogs in the country, we find sixty thousand hogs in Chicago in one day and the price dropped thirty cents per hundred— "a ruinously low price," caused by "ruinous competition" among hog sellers.

The wheat crop of 1911 was one of the shortest in years, and yet the wheat sellers hurried their winter wheat to market and put seventy million bushels on the market at once, and the price went 20 cents a bushel below the natural supply and demand price, "a ruinously low price," caused by "ruinous competition" and trust-owned farm papers, told the farmers that seventy and eighty cents was the market price, and discouraged them in every way possible from trying to get together and prevent low, unjust prices caused by "ruinous competition."

One half million dollars worth of apples rotted in our orchards in 1911, simply because farmers glutted their little home markets by "ruinous competition," and brought the price so low they were not worth picking up, while consumers in many localities paid high prices. The dry season in 1911 caused a very short potato crop, and thousands of farmers being compelled to buy, paid over one dollar per bushel while in the best potato sections, potato growers sold their potatoes for thirty cents per bushel.

Competition among sellers is the cause of this condition. As long as enough farmers will supply the market with any product at cost or below cost of production, others have no protection. Our only hope is in organization and co-operation to prevent "ruinously low prices." Everybody except the Indians, idiots and farmers are organized, hence we are a mob in the business world, contending with well organized forces. One thousand well organized soldiers can whip a mob of ten thousand. Our power when united is unlimited because we produce the necessaries of life. We feed and clothe the world. Our products are so valuable that we would cause great suffering if we should stop marketing for ten days all over the United States.

We do not believe in the trust principle of trust co-operation of the few to the sorrow of the many. Co-operation, which enables a few brainy, selfish men, to take hundreds of millions of dollars produced by farmers and wage-earners, is the great curse of our country. They rob us both as producers and consumers. They take and center more and more of the wealth-makers' wealth in the hands of the wealth takers. Seventy million of our people must labor early and late, and subsist on the mere necessities of life, that the other thirty millions may live in ease and luxury. The Farmers' Equity Union plan of fraternal, golden rule co-operation, will solve this great economic question.

The seven million wage-earners must form great industrial unions which promote the intelligence, morality and fraternalism of their members and educate them to be true blue golden rule co-opreators. Gradually they will be induced to take stock in the mines, factories and railroads until they hold the controlling interest. They will then get all the wealth they produce. Their manhood, intelligence and character will be developed until they will be a very different class of people. This should be the aim and purpose of all labor unions, and strikes will be finally a thing of the past. Millions of farmers must be united in the Farmers' Equity Union. Our principle is the greatest good to the greatest number. Equal rights to all is our fundamental doctrine. Our exchanges are self-supporting as they run on a safe, sane business plan. Study our proposition in the Equity Text Book. This Union will prevent "ruinously low prices" on all farm produce. We will not build up a visible supply of seventy million bushels of wheat, as a club to hammer our price down, so that the speculators can take millions of dollars which we produce. We will put our hogs into our own stock yards, and finally through our own packing plants, and regulate the supply to the demand. We will market our onions, cabbage, potatoes, apples ets., through our own exchanges, paying producers a good price and at the same time making consumers' prices more reasonable.

At every good shipping point we will reverse the present conditions. Instead of paying as little as possible, our markets will pay as much as possible for all farm produce and we will finally control shipments to central markets so as to prevent "ruinously low prices" there. Brother farmers, let us keep in mind that the success of this movement depends largely upon education, a slow but sure process.

The fundamental principles of Equity Union are as follows:

1. Allow no man to take stock in our exchange unless he is a member of the Equity Union.

2. Never allow the directors to declare over five percent dividend on the stock subscribed.

3. Never pay outsiders any of the profits, nor give them any benefits, but always hold the door open for them to come in.

4. Do not allow business men to take stock in our exchanges unless they come in as land owners, or in case we start co-operative stores and take them in as managers.

5. Allow no man to own over four shares, unless it is absolutely necessary.

6. Allow no member to draw out any cash nor interest until he has four shares.

7. Build up the capital till every member has exactly one hundred dollars in the exchange.

Go to every meeting and see that these principles are not violated or changed. Wherever they are being adhered to, farmers are making a grand success in the Equity Union. Get every farmer in your community to read the weekly paper, for whoever reads the paper every week, will become a member and stick to the Union. We want educated co-operators, to carry out the principles of our great organization.

The most of our corporations are now owned by a few "money kings," but when the Equity idea is fully carried out, in our country, the people will own the corporations, co-operatively, through industrial unions and the individual or man will be the unit and not the dollar. "Uncle Sam," will be representative of the industrial unions and the umpire regulating and controlling all co-operative corporations. Then the best principles advocated by far seeing lovers of their fellowmen will be in practical operation in our country, and its people will be the happiest and most prosperous on this earth.

RESPONSIBILITY FOR BAD CONDITIONS

Logan Waller Page, director of the United States office of public roads, in an interview given out in Washington, said: "Philosophers who have been deploring the trend of population from country to city might as well save their energy unless they are prepared to help change the conditions, responsible for the migration. At the root of this condition are the poorly-kept roads of the country. It is certain that the farmers do not get the use of their shares of the money earned in the United States."

"There are now over ninety million people in this country, and nearly one-third are the farmers and their families. The products of the farm are responsible for more than one-third of the wealth and commerce in the country. No one can say, however, that one-third of this wealth is used by the farmers, in the betterment of the country districts. **It is due to the inactivity and lack of co-operation among the farmers, that country districts have become depopulated.**"

"The 'Back to the Country' movement has had a considerable vogue at sumptuous banquets in cities, but the movement has not reached far out of town, for the simple reason that life in the slums, despite all theories, frequently is much more livable than life in the country. While fine phrases can be made in the discussion of the joy and health of hard labor in the open air and under the sun, it is a fact that conditions in the cities are much more healthful today than conditions in the country. Sanitation is better, and the air in tenements is little worse than the air in many country houses where, because of the extreme cold and lack of proper heating apparatus, the windows must be kept closed from October until May. Improper sanitation, poor drainage and poor highways in the country have brought about as great a percentage of diseases as exist in the cities. While country people are free, as a rule, from tuberculosis, they are extremely liable to typhoid fever, pneumonia and other diseases that are brought on buy improper drainage and improper ventilation.

"The work that is being done toward the improvement of roads throughout the country will change this condition. Improved roads will give the country districts the improvements enjoyed by fashionable suburbs, and will improve drainage and wipe out isolation. In most localities life on farms invariably becomes, as a result of bottomless, roads, isolated and debarred of social enjoyment and pleasures, and country people in some communities suffer such disadvantages that ambition is checked, energy weakened, and industry paralized.

"Under such conditions it is but natural that persons engaged in farming, especially the younger folks, should seek the life and gayety of cities. There they feel they will find recreation, variety, youth, beauty and music. The difference between good and bad roads is often equivalent to the difference between profit and loss. Money wisely expended for this purpose is sure to return tenfold.

"I quote the entire interview, as it should all be read by my country people, but I call attention especially to one sentence, viz: "It is due to the inactivity and lack of co-operation among farmers that country districts have become depopulated."

Here is a truth that the Farmers' Equity Union is making strenuous efforts to bring to the mind of every farmer and his family. The conditions in the country must be, and are being improved by the farmers themselves, through organization and co-operation.

Our local union members find many ways to co-operate for the uplift of their own community. In some localities where roads are likely to become bottomless or impassable, the split log brigade gets busy, every member doing his part and every road in the neighborhood receiving ample attention at the right time. One local union president has the various country schools around his town to furnish a program of songs and speeches for the Equity Union at the regular monthly meeting, the first Saturday of each month. Every community around the town becomes interested in Equity Day, the first Saturday of each month. Sociability and fraternalism are cultivated between all neighborhoods surrounding each marketing place. Spelling bees, declamation contests, public debates, corn shows, musicals and baseball matches all serve their purpose to bring the farmers and their families together, and make life in the country more attractive and more endurable.

The first Saturday of each month is "Equity Union Day," and must be religiously observed by Equity Union families all over the United States. Farmers must be members of a National Union, meet regularly and contribute money conscientiously and be active workers. Inactivity, carelessness, neglect and indifference along this line must be overcome. We must work constantly and faithfully for members, meetings and money, all of which are absolutely necessary to the success of a National Union of farmers. The more co-operation among the members, the more benefits will be shown and the more growth there will be of the organization. We must co-operate for the improvement of our roads, our seed and our stock. These old shacks, called school houses, must be replaced with modern buildings, with a furnace underneath, a good bell on top and modern equipments inside. The term in the country should be eight or nine months annually. The country children are worthy of the same advantages, comforts and conveniences as city children, and often appreciate them more and make better use of them. A want of organization and want of co-operation among farmers is responsible for the bad conditions in the country wherever they exist.

The Farmers' Equity Union presents a practical business plan of co-operation, which must be carried out by its members more fully and completely each year, as the people are educated and become able. We must combine, concentrate and co-operate. Every farmer must be brought into line and be educated to be a Golden Rule Co-operator. It will make him a better husband and father, a better citizen and a better business man. To be a true co-operator, he must be persuaded to overcome his selfishness, narrowness and ignorance. He must be a reading, thinking man. He must care for others as well as for himself. He must be fraternal, public-spirited and generous hearted. He must be an active worker for co-operation in his community, state and nation, and for the co-operation of the many, for the benefit of all. This will defend us and protect us against the co-operation of the few to the sorrow of the many. It will bring wealth producers into their own. The wealth produced in our country districts will no longer be centered in the hands of a few millionaires.

It will be brought out into the country for good roads, beautiful, comfortable homes, modern school houses and country churches. We must work for the education of the individual farmer and his family. Lift up the individual and you lift up the masses. Make the individual farmer intelligent, moral and fraternal and he becomes a true blue co-operator, and is no longer a clog or brake in the wheels of progress.

The campaign of organization and education must be incessant. More honest, sincere educators must be kept in the field agitating, educating. More and more literature must be printed and circulated. Every member is required to be a subscriber for our paper. More agricultural papers must be induced or persuaded to allow space for the discussion of economic questions. The intelligent distribution of our valuable crops is of equal importance with their production.

Let me repeat that inactivity and indifference in regard to organization, and hence a lack of co-operation among farmers, is responsible for bad conditions on the farms. It is responsible for low, unjust prices on our largest crops of finest quality. Responsible for high extortionate prices on what we buy. Millions of bushels of winter wheat left the farmers hands in 1914 at 70 cents a bushel, for which speculators received over $1 per bushel. Millions of dollars were lost that year by farmers by not being organized.

As an organized body farmers have all necessary power of defense and protection. No need to submit to a single injustice or wrong when organized right. The farmer who will not unite is responsible for innumerable wrongs against his business, his home and all country people. Farmers must be continually stirred, aroused, interested, educated. He who would be free must himself the blow strike.

In states where the Equity Union is being pushed the farmers are awakening as never before and becoming active workers for co-operation. They are meeting the first Saturday of every month, supporting the union by regular dues and discussing ways and means for the promotion of the general welfare. They are gradually educating their community to be co-operators, and building up a co-operative business worth thousands of dollars annually to the farmers around their town. The few hundred dollars they spend for promotion work and for the support of Equity Union comes back in thousands of dollars annually, and they wonder why they have been so short-sighted, indifferent and inactive in the past.

The fifty new local unions started in North and South Dakota, right in the dry section, are all growing and will soon result in an Equity Exchange in every union. Enough wealth is produced in every farming community each decade to tide over comfortably the short years and failures of crops, and to bring blessings, comforts and luxuries enough to hold our boys and girls in the country. But inaction, neglect and indifference to organization and lack of co-operation among farmers allows a large per cent of the wealth they produce to drift away from the country into the hands of millionaires in our large commercial centers. Brother farmers, through organization and co-operation we can come into our own. The farmer who neglects or refuses to unite or who joins a union and neglects to attend meetings and pay his regular dues, is responsible for bad conditions in the country.

If two hundred farmers will unite in a local union, at any good town, put twenty-five dollars each into an Equity Exchange, erect an elevator for co-operative buying and selling, and carry out our practical plan as explained in our Equity Text Book, they will in a few years have a capital of twenty thousand dollars invested and banked. They will be equipped to handle all of their produce at actual cost, and also coal, lumber, flour, feed, salt, fencing, wagons and all farm machinery, and when failures come in succession as in the Nortewest in 1910 and 1911, each locality will be in a position to furnish themselves the very best of seed for the next season. They will have financial standing in the business world, which will carry them through the bad season. When thousands of Equity Exchanges are linked together in the great Equity Union chain, they will have their own grain company, with a branch in every central market; they will hold the surplus of all crops in the country, regulate the supply to the demand and maintain a fair, equitable price on every product and every valuable crop, and at the same time consumers' prices will be more reasonable.

Thousands of Equity Exchanges will center their trade with individual factories, mills and mines, and reduce prices fifty per cent. Inactivity, carelessness and negligence of organization and lack of co-operation among farmers is the prolific cause of bad conditions on the millions of farms in our rich, prosperous country.

THE FARMER A BUSINESS MAN

The Farmers' Equity Union is educating the farmers to follow business methods and principles, which are considered just and wise in the business world. Every business man has a place of storage for his goods. Equity Union is urging farmers more and more everywhere, to erect warehouses for storing cotton, potatoes, beans, etc., and granaries and elevators in the country for the storage of grain.

Every merchant stores his goods and holds them for a profitable price, not as a speculator but as a business man following business methods. Equity Union is schooling farmers in the same wise business principles. When farmers are thoroughly organized and educated to store their goods in the country, and have a minimum price for the central market, our prices will be as steady and sure as the price of anything we buy.

The imperative demand for our valuable products will call for a steady stream of them, to every central market, every day of the week, and at the end of each year we will find the entire crop was needed at a just price to the producer, and that consumers have also been protected from grafters, speculators and extortioners.

The business man is fully informed on the business side of his calling. He spends more time reading, studying and planning than he does working with his body. The time has come when farmers must use brains as well as muscle and they must also be educated to be golden rule co-operators. They must be educated to buy and sell co-operatively so that their produce will reach the central market at actual cost of handling and bring a price that is just to them as producers. This they will never do as a mob of sellers, with no equipments for storing and handling their crops.

In the business world the seller sets the price. It has often been said that there are only three classes in our country, who do not set a price on what they sell, viz., Indians, idiots and the farmers. Equity Union will take our farmers out of that class and place them where they belong, viz., in the business world, following business methods. One great power which business men have to secure profitable prices is the fact that they have a place of storage, hold their goods for a profitable price and get the price by regulating the supply to the demand. They do not put too much on the market at once. They supply the demand that comes each day at their price, and then stop selling till the next day. From day to day they regulate the demand, and thus maintain a steady equitable price.

When the farmers have a good wheat crop, or any other grain crop, they put hundreds of car loads into our central markets and often break their prices thirty per cent at threshing time, and then keep so much piled up under the noses of big buyers, till the next crop comes so that they hold the price down on themselves on the entire crop, and lose millions of dollars by not being organized and equipped to regulate the supply to the demand, as business men do. Bradstreet showed that by December 1st, 1902, seventy five per cent of the wheat crop had left the farmers' hands. They put a twelve months' supply on the market in five months, and by so doing the farmers lost one hundred million dollars, while the railroads, millers and speculators made hundreds of millions of dollars out of the crop, and the consumers paid a good stiff price for flour.

Low prices in our central markets are caused by the way that farmers market. They sell as a mob and have no business method to follow. When the price begins to drop many of them begin to haul. When they should all stop, they haul more. We know that one thousand soldiers can whip a mob of ten thousand, because they are organized and equipped. Farmers must be organized, educated and equipped to follow business methods. They are losing one hundred times more each year, by not being organized, than the organization and full equipments for co-operative buying and selling would cost them.

There are hundreds of places in our country where the farmers contribute enough patronage each year to build a fine, large elevator, fully equipped to handle at cost all of their grain, hay, stock, flour, feed, fencing, fertilizer, twine, tiling, salt, wagons, and all farm machinery. They pay for a big elevator for a millionaire owner in some city, and let him own it and make them pay for it again and again every time they have a good crop. Farmers, is this not the height of folly? Will it not pay to go together in a local union of the Farmers' Equity Union, get a hundred members, organize a stock company, build an up-to-date elevator, buy and sell on a safe margin and prorate back all profits till you have all your money you put in the elevator back in your pockets and you stop paying for the millionaires' elevators over and over again?

Read the Equity Text Book carefully and get a clear full understanding of this plan of co-operation. Then get five farmers or more to join you in a local union and the national president will come and build you up till you have one hundred members and a good elevator or Equity Exchange. Our plan if carried out will finally unite ninety per cent of the farmers around your town and keep them united.

Another power of the business man to obtain profitable prices is this, viz., by storing his goods and holding them for a profitable price, he forces the buyers to come to him and this gives him the power, instead of the buyer, to set the price. Farmers are in the habit of running to buyers and asking them the price. When organized we will turn the tables, and the man who tells us our wheat crop will not be called for at a fair price in all of our leading markets, might as well assert that our one hundred million people will stop eating bread. To claim that our hogs will not be called for at a fair equitable price to the farmer is as foolish as to assert that the people will eat no more meat.

It is right for every business man to set a price that is fair and equitable, and it is equally as just and important for farmers to employ the same business method, that they may have the same power to price their products. The greatest power of business men to obtain profitable prices is in the fact that they all refuse to sell below a just price, or in other words, they stand together for the price. The Farmers' Equity Union is organizing and educating with this great purpose in view for farmers. We are not organizing however to put high prices upon consumers, but we believe we can grind our wheat into flour ourselves, and sell it to the consumers cheaper than they are now receiving it.

We are spreading this union from Ohio to Montana, and from North Dakota down to Texas. The Grange is growing stronger every year in the east. The farmers' educational co-operative union is strong in the south and slowly spreading in some other states. The two thousand farmers' elevator companies now have a state association in seven states. These great educating forces are at work from ocean to ocean and from the gulf clear up into Canada.

Farmers, do not be afraid that you will join a losing cause, when you join the Equity Union. We are sure to win. No farmer should sulk in his tent, saying: "It can't be done, they won't stick." Come in and stick yourself. Every farmer who refuses to unite is responsible for bad conditions and low prices. A strong union is a protection and is a guarantee against low, unjust prices on what we sell and high extortionate prices on what we buy.

A MINIMUM PRICE A PROTECTION

The protection we farmers need is a union that leads every farmer to have a minimum price on grain, hay, stock, wool, cotton, and every product we sell. This trusting to luck or to speculators and trusts to price our ten billion dollars worth of products each year is the heighth of folly, and is the cause of the loss of millions of dollars on every good crop the farmers raise. Every business man has a minumum price below which he cannot afford to sell, and below which he will not sell and does not have to sell. He is protected against a low, unjust price by having a minumum price, and by the fact that every other dealer has nearly the same minumum price, or in other words, stands with him for the minumum price. So farmers ought to have a minumum price for every product. There is a minumum price below which we cannot afford to sell wheat, corn, oats, barley, wool, cotton, hay, hogs and cattle.

The farmers should be organized so that they would never allow good wheat to sell in our central markets or anywhere else at a low unjust price. Hogs fed on fifty cent corn ought to bring six cents a pound. If the natural law of supply and demand gives us more, I suppose we have a right to take it. Manufacturers, railroaders, wholesalers, jobbers and retailers all protect themselves against ruinous prices by having a minumum price, below which they will not and do not have to sell. In the business world the sellers set the price. The individual farmer has no power, whatever, to price his products, and he has no power to defend himself against a low, unjust price. He has no power whatever, single handed, to defend himself against combinations and manipulations in our central markets. He, as an individual, has no way to protect himself against low, unjust prices caused by the dumping system.

He may thresh ten thousand bushels of wheat, of finest quality, store it safely, have it insured and price it at one dollar a bushel, which is low enough for good wheat, but if the central markets are fed regularly the entire year at eighty or eighty-five cents a bushel by the mob of sellers, he simply holds the bag for the others. His only hope is an Equity Union, national in scope, on a practical plan of co-operation. He must lay aside his prejudice and suspicion, and join the union, get others to join, go to every meeting, and be willing to support a union with one dollar per year. He must also read the Equity Union Exchange every week and read carefully the Equity Text Book.

Our members must meet regularly every month. We must make it our religious duty to quit work one-half day every month and go to an Equity Union meeting, where we shake hands with our brother farmers, exchange ideas on subjects in which we are directly and mutually interested as farmers, become better posted on crop and market conditions and have a definite understanding on the minumum price of all products. These are simple, easy things to do, which every farmer can do if he will, but when one million farmers and their families do these simple easy things regularly, and faithfully every month, we will have the most powerful industrial union in our country, which will insure every member against low unjust prices, and we will dominate every political party in our land. We must keep in mind constantly, that economic freedom is only possible to an intelligent fraternal people. Our union is making strenuous efforts to promote the intelligence, morality and fraternalism of its members by declamation contests, corn shows and lectures on good roads, and improved schools, but our goal is co-operation in buying and selling on a practical plan that will defend us and protect us against low, unjust prices on what we sell, and high, extortionate prices on what we buy. We want an Equity Exchange at every good shipping point and a co-operative grain company with a branch in every large city.

This will prevent the glutting of central markets and prevent breaking down our minumum price on all products. This organization will be worth at least, fifty dollars per year to the average farmer when its minumum prices are maintained and our co-operative plan of buying and selling is fully carried out. It will be worth two hundred and fifty million dollars annually to the five million plow-handle farmers now renters, on mortgaged homes or on claims on the frontier, enduring the hardships of frontier life.

Ex-president Roosevelt, in his message to congress accompanying his report of the commission on country life, said: "If country life is to become what it should be, and what I believe it ultimately will be—one of the most dignified, desirable and sought-after ways of earning a living—the farmer must take advantage not only of the agricultural knowledge which is at his disposal, but of the methods which have raised and continue to raise the standards of living and of intelligence in other callings. Those engaged in all other industrial and commercial callings, have found it necessary under modern economic conditions, to organize themselves for mutual advantage and for the protection of their own particular interests in relation to other interests. The farmers of every progressive European country have realized this essential fact, and have found in the co-operative system exactly the form of business combination they need.

"Agriculture is not commercially as profitable as it is entitled to be, for the labor and energy which the farmers expend and the risks that they assume, and the social conditions in the open country are far short of possibilities."

The first great, general and immediate need is co-operation among farmers to put them on a level with the organized interests with which thy do business. Senator McCumber of North Dakota, in an address to a large gathering of farmers at Fargo, said: "There is a remedy for all the inequalities and injustices mentioned in this argument. The instrument of defense has for years been in our hands. We have been as soldiers sleeping by our loaded guns, while the enemy has plundered our fair domain. If the injuries, the inequalities, the injustices we have suffered have been the result of organization or combination of the other industries, then it needs no great amount of thinking to see that our remedy is a counter organization."

THE NEBRASKA CO-OPERATIVE SYSTEM

During the past winter, Nebraska has taken front rank among the states in its liberality toward co-operative companies. Hitherto such concerns had only the ordinary corporation law under which to organize, and in some ways it has been found unsuited to co-operative theories and practices. Since the old corporation law was the only one in the statutes, the practices of co-operative concerns have been restricted by unfriendly court decisions, until many so-called co-operative companies are such in name only, all their methods following closely the well-beaten corporation path. Nebraska co-operators have, however, broken the precedents of generations and have blazed out a new path that doubtless will be lengthened and broadened hereafter by co-operators in other states until this class of capitalists will enjoy the universal recognition to which their enterprise so entitles them.

The new law clearly defines a "co-operative" corporation or company as one that authorizes the distribution of its earnings in part, or wholly, on the basis of, or in proportion to, the amount of property bought from or sold to members, or of labor performed, or other service rendered to the corporation. This does not mean that the company "must" distribute its earnings in proportion to business from the stockholders, but it opens the way so that when a co-operative elevator company has an exceptionally successful season, with large volume of business and exemption from losses, and finds a profit sum in its treasury equal to 30 or 40 or 60 per cent of its capital, it may pay to all stockholders alike a reasonable interest for the use of the money and then distribute the excess earnings "in proportion to the business each stockholder has furnished the company." This will put the excess profits back again on the farms from whence they came, increasing the prosperity of the farmers and indirectly benefiting every class of business in the community.

Another feature of the new law is that it expressly places in the hands of co-operative companies the power to choose who shall own stock therein. Heretofore courts have been deciding that any one had a right to buy a share of stock wherever he could pick it up, and compel the company to transfer it on the books and recognize the buyer as a member of the corporation. This had resulted in enemies buying stock from parties moving away or where from some other reason shares were found for sale. The new law expressly confers on a co-operative company the power "to make by-laws for the management of its affairs, and to provide therein the terms and limitations of stock ownership and for the distribution of its earnings."

There is still another feature of co-operation that Nebraskans have been able to secure in spite of adverse laws. Co-operators frequently insist on equality of voting power in stockholders' meetings, regardless of the number of shares held. Nebraska laws strictly forbid any interference with the right of stockholders to vote in proportion to the stock owned, and this provision is engrafted in the state constitution. Nebraska co-operators have found a way, however, to capitalize their business and secure the desired single vote for each stockholder. This is accomplished by a provision of the by-laws permitting a person to buy or own only one share, and while the practice has heretofore been carried on in disregard of law and by a sort of "Gentleman's agreement," the new law makes all such by-laws legal, and the necessary capital can be raised for any co-operative enterprise by making the value of the shares high enough so that when all interested persons have one share, there will be enough capital for the desired purpose. In practice among farmers' elevator companies it has been found that shares of $100, each and restricting the ownership to one share per person, usually produces the most satisfactory results.

However, the Equity Union makes it easy for the poorest renter to get the benefit of co-operation by placing the shares down to $25 each. If he is a patron, enough will accrue to his credit in a few years to give him three more shares. He can draw nothing out till he has four shares, the limit. Every man who takes only one share builds our capital up $75 by his patronage. When the Equity plan is fully developed, each stockholder has four shares or one hundred dollars in the company. Then all share-holders are equal. Undesirable stockholders are kept out by limiting the shares to four to each stockholder, and by not declaring over 5 per cent dividends on the stock subscribed. Buy and sell on the same margins as competitors and prorate to stockholders all over 5 per cent on stock subscribed, and your company will live, grow and succeed. You count out cash to each stockholder once each year, which is a practical demonstration to him of the advantage of co-operation over the old capitalistic idea of one man pocketing all the profits. Outsiders will finally be convinced by these practical demonstrations and unite. We can unite 90 per cent of the farmers at each shipping point on our just, co-operative plan. Every state ought to have the new Nebraska co-operative law. Read it carefully:

SENATE FILE NO. 88.

An act to define co-operative associations and to authorize their incorporation, and to declare an emergency.

Be it enacted by the Legislature of the State of Nebraska:

Section 1. For the purpose of this act, the words "co-operative company, corporation or association" are defined to mean a company, corporation or association which authorizes the distribution of its earnings in part, or wholly, on the basis of, or in proportion to, the amount of property bought from or sold to members, or of labor performed, or other services rendered to the corporation: Provided, that nothing in this act shall be construed as in any way conflicting with or repealing any law relating to building and loan associations or installment investment companies.

Sec. 2. Any number of persons, not less than 25, may be associated and incorporated for the co-operative transaction of any lawful business, including the construction of canals, railways, irrigation ditches, bridges, and other works of internal improvements.

Sec. 3 Every co-operative corporation as such has power: First—to have succession by its corporate name.. Second—to sue and to be sued, to complain and defend in courts of law and equity. Third—to make and use a common seal, and alter same at pleasure. Fourth—to hold personal estate, and all such real estate as may be necessary for the legitimate business of the corporation. Fifth—to regulate and limit the right of stockholders to transfer their stock. Sixth—to appoint such subordinate officers and agents as the business of the corporation shall require, and to allow them suitable compensation therefor. Seventh—to make by-laws for the management of its affairs, and to provide therein the terms and limitations of stock ownership, and for the distribution of its earnings.

Sec. 4. The powers enumerated in the preceding section shall vest in every co-operative corporation in this state, whether the same be formed without, or by legislative enactment, although they may not be specified in its charter or in its articles of association.

Sec. 5. The fees for the incorporation of co-operative corporations or associations shall be the same amounts as those provided for like capitalization of general corporations in the state of Nebraska as provided in Section 5905 of the Compiled Statutes of Nebraska for 1909: Provided, that any co-operative corporation or association, being such under the definition given in section (1) of this act is hereby authorized to file with the secretary stating that it is a co-operative corporation or association as above defined, and from and after the filing of such declaration with the Secretary of State, it shall be entitled to the same legal recognition as though its articles of incorporation had been originally filed under this act, and the fee for filing such declaration shall be Two Dollars, subject, however, to the general incorporation laws of the state except as herein modified and changed.

Sec. 6. Wheras, there being an emergency, this act shall take effect and be in force from and after its passage.

THE GET-TOGETHER AGE

As the physical world is bound together by bands of steel, so the mental world is bound together by electricity. Go to the telephone, get the long distance, and you find there an electric current that will carry your voice a thousand miles, right to your friend's ear, so plainly and so distinctly that he will understand every word you say.

You can stand on the shore of one ocean and communicate instantly with your friend on the other side. A new invention now enables eight persons to talk over one wire without confusion. Electricity has given our mind wings. We are no longer on the ox-team basis, but on the wireless basis. Steam and electricity have drawn the business world irresistibly together. Modern education has also developed independent, logical, keen thinkers never dreamed of in former years.

Steam, electricity and modern education have brought the business world naturally and irresistibly together and given rise to "big business." As a natural result we have enormous corporations with billions of capital engineered by a few brainy men. For our government to attempt to destroy these corporations would be to attempt to turn the wheels of progress backward and to return to the days of sailing vessels, ox teams and stage coaches. Let us face the fact that this is the electric age which is bringing the business world, naturally, irresistibly together. Let us realize the truth that the combinations grow larger and stronger every year in the business world.

Our danger, however, is not in the size of our corporations but in the greed and selfishness of a few brainy men who now control them. We need have no fear of combination and co-operation when based on right principles.

We want federal regulation and not destruction of our corporations. But the final solution of the problem will be true co-operation among all the people. Industrial unions will manage the business and control the politics of this country. Industrial unions will educate the so-called common people (God's kings and queens) so they will own co-operatively the corporations and run them on the principle of equal rights, opportunity and protection to all. I do not believe in government ownership of all property. It would not settle the strife between labor and capital. Paternalism will never develop a self-reliant, free, independent people, and will eventually destroy individual initiative. But industrial unions which continually promote the intelligence, morality and fraternalism of their members will educate the people to be true co-operators and they will finally own and control every big corporation in our country co-operatively. When this is true we need not fear of the size of our corporations. They will grow larger instead of smaller. The coal miners' union must promote the intelligence, morality and fraternalism of its half million members, until the miners understand thoroughly true co-operation, take stock in the mines and own the major part of the stock. Then the intelligent, moral majority will control the one gigantic coal corporation of our country. They will elect their own bosses, and be obedient to rule by the majority. All the wealth they produce will go to them, and there will be no coal barons to rob them. Each individual will feel that he is a man with every right of manhood guaranteed.

Our railroad then will be one gigantic corporation, owned, conerolled and run by the millions of men and women who operate them. Then more care and economy will be exercised by all and greater efficiency rendered to the public. The principle of true co-operation will be carried out by the intelligent, moral, fraternal majority. The railroad magnate will be one of their own number, employed by them, instead of the selfish, domineering millionaires, and strikes and paralysis of business will be a thing of the past.

The millions of people in our factories must be organized into industrial unions and educated to be true co-operators. They must own and control the great factories in which they labor and run them on true co-operative principles.

The Farmers' Equity Union is one of the pioneers in this great movement for economic freedom of the people. We come to the farmers with a clear, well-defined, practical plan of organization and co-operation. We emphasize the educational feature more than anything else. Our constitution says our chief object is the promotion of the intelligence, morality and fraternalism of its members. We believe that economic freedom is only possible to an intelligent, moral, fraternal people. We insist on a continual campaign of education by literature, a weekly paper, lectures and meetings. Everything possible must be done to make each meeting attractive, interesting and instructive. Farmers must get together. They must be more fraternal. Lay aside prejudice, suspicion and everything which separates them. Farmers, this is the get-together age. Two hundred or more farmers around each good town must unite in a local union, organize an Equity Exchange, build a farmers' elevator, buy and sell co-operatively on our plan, and they will show benefits to members which will keep them united. As you read this determine now to start a local union at your town. Send for our Equity Text Book which explains fully our plan of organization and co-operation.

We want one thousand strong links made at one thousand good towns. Then we can reduce the price of farm machinery 50 per cent to our members and also prevent gluts of our central markets, which often lower our prices 30 per cent below what the natural law would give us.

Farmers, let us wake up to the fact that this is the get-together age, and the class which refuses to unite will fall far behind the procession with the ox team brigade. The farmers have paid for many thousands of elevators for the grain trust, and would be astounded if they knew how many times they have paid for some of them. The Equity Union is persuading them to pay for their own elevators by conserving their patronage on co-operative principles, and build up their capital in the same way until they have fifteen or twenty thousand dollars in each Equity Exchange. The money each member pays into the Exchange is simply a loan to his elevator, and comes back in cash as soon as sufficient patronage has been given the Exchange. If 200 farmers will get together and put twenty-five dollars each in an Exchange and carry out our plan fully, they will have a capital of twenty thousand dollars in a few years, which is built up out of the patronage they now give away, and each year afterward they can divide a five thousand dollar melon among themselves instead of giving it away to unnecessary middle men by giving away their patronage.

When one thousand of these links are on our plan, each with a good warehouse and financial standing in the business world, the National Union can take the entire output of an independent coal mine, of a flour mill, a cordage company, a wagon factory or machine factory of any kind.

We save the cost of advertising and of traveling men, and enable each factory or mill to run steadily the entire year. We take away all risks of sale. We pay only for actual cost of material and labor, and a fair profit guaranteed to the factory. On this plan we come to the factory with our patronage organized and reduce the cost 50 per cent to our members. A difference of five or ten dollars on every wagon and machine between members and non-members will unite the farmers solidly. A solid union of farmers will lead to more fraternalism, more and more brotherhood and more economic freedom. This is the get-together age, and farmers must learn it.

THE NEW DECLARATION OF INDEPENDENCE.

As I am writing this on Independence Day, I wish to speak of the new declaration of independence being made by the American farmer. The badge of the Farmers' Equity Union is a shield containing the flag of our country, and is a badge of freedom. Every farmer who dons this badge makes a new declaration of independence. He is taking a brave stand for economic freedom. Freedom from trust rule and domineering "Captains of Industry." He is joining an Industrial Union which is the champion and brave defender of his every economic and political and religious rights.

The millions of farmers of our country are our greatest consuming class. We buy and sell more than any other class. Nine or ten billion dollars' worth of real wealth brought into actual existence each year is the boast of the American farmers, and well they may feel proud of the achievement, as it is all wrought out by the honest labor of the toiling yeomanry. This is an immense sum. It means the actual earnings of about one-third of our population. It is produced by millions of people who live on our farms, much of it by wife labor and child labor. None of it is produced without actual toil, care and anxiety. If you want to obey fully the Divine command, "In the sweat of thy face shalt thou eat bread," go out on the farm for ten years.

The farmers' products are not only more valuable but more important than of all other classes combined. We feed and clothe the world. We furnish the very necessaries of life without which the race could not exist. The farmers' business is fundamental. Every other business depends on the success of the farmer. Hence we see continual solicitude for the successful growth and care of crops on the part of the business world. We see railroaders, bankers, merchants and manufacturers all interested in educating this industrial slave how to farm better, how to produce more, how to raise bumper crops.

Millions of dollars are spent annually for agricultural papers, farmers' institutes, experiment stations and agricultural colleges, all teaching us how to produce more, which is all right. We are not surprised at the interest show by railroaders, bankers, merchants and millers in our business, because we know they get billions of the wealth we produce, especially when we raise maximum crops. We would not have less intelligence in farming, but more of it. The carrying out of our plan means a farmers' institute in every good town the first Saturday of every month, not only for the purpose of hearing lectures, but to discuss among ourselves the various problems of life. Bue we do insist that the great mob of farmers are the industrial slaves of the organized classes in the business world, and always will be till they are thoroughly organized and educated to buy and sell co-operatively.

Their nine or ten billion dollars of wealth produced annually is sliced down on every side by organized buyers and sellers until the farmers often find themselves reduced to a bare existence, and their boys and girls fleeing from the condition in the country to the cities, where the country's billions are centered.

Millions of farmers' families are denied the common comforts of life. They must continually skimp, save and economize. They are honest and industrious, but are robbed continually by our present industrial system which takes the wealth from their toiling hands and places it in the unholy coffers of the greedy, selfish rich. Brother farmers, here is our greatest difficulty: we stand the failure of crops, but when we succeed in producing a big crop of fine quality, in our unorganized condition, we play right into the hands of speculators and trusts, and get less for our finest crops than for the small inferior crops. We sacrifice our best crops of cotton, wool, grain, stock and all produce by competing for the markets where we should co-operate. We feed and fatten speculators and trusts because we are a mob of sellers, and they are thoroughly organized.

We support by our patronage millions of unnecessary middle men and their families in luxury, for whom we have no earthly use. We allow the sugar trust, salt trust, match trust, oil trust, furniture trust, millers' trust, shoe trust, clothing trust, lumber trust, machine trust, grain trust, cotton trust, wool trust, meat trust and milk trust to slice that nine billion dollars down on every side year after year, and the result is millions of farmers slaving and saving for a lifetime to make millionaires of others.

If the farmers received the wealth they actually produce each year and would use it right, every agricultural township would have a high school and small experiment station in its center. Good, hard roads would lead from every farm to schools, churches and towns. Beautiful homes and churches in the country would be as common as in the towns. Hundreds of old shacks now called school houses, would be replaced by neat, comfortable buildings, with a furnace underneath and a good bell on top. Our boys and girls would be attracted to country life with its healthy atmosphere for the physical, mental and moral man. Thousands of middle men, now parasites, would be compelled to become producers instead of leeches.

When the great mob of farmers becomes a well-organized body in the Farmers' Equity Union and adopt our practical plan of co-operation, we will prevent the wealth we produce in the country from drifting away from us into the large commercial centers. We will bring it out into the country to bless the people who toil for it, and our country people will enjoy the full fruit of their labor. How will we unite them? Throw away the cowardly word "can't." Organize, educate and persuade them to strike for freedom. He who would be free, must himself strike the blow. Every farmer must wear the little blue badge, the symbol of fraternalism and freedom. He must sign the new declaration of independence. He must swear allegiance to the Equity Union through which he will win economic freedom. He must be for all the other brother farmers that they may be for him. He must loyally support his union with his influence, his money, time and labor. One-half day out of every month spent in an Equity Union meeting by one million farmers and their families, discussing the production, distribution and marketing of crops would be worth as much as twenty-five days each month spent in the fields. This is not a radical statement. It is true.

If the wheat growers of our country had done this in 1914 they would have received two hundred million dollars more for that crop, and consumers would have paid no more. If the wheat growers of the United States were meeting now in Equity Union meetings every Saturday afternoon, wheat would go to one dollar a bushel in ten days and remain there till November, when it would go to $1.05. In January it would go higher, in March a little higher, and a just price would be received for the entire 1914 crop. If the hog men had been thoroughly organized in 1907 and 1908, and as a result met regularly once or twice each month all over the country, and have held big central meetings in Chicago, St. Louis, Kansas City, Omaha, Sioux City and Soiux Falls, they would have sold those two fine crops of hogs at a profit instead of a loss, and there would have been no shortage in 1909 and 1910 and consequent high, extortionate prices to consumers. The farmer who refuses to don the little blue badge, sign the new declaration of independence and neglects to attend the Equity Union meetings and contribute $1 annually is not as patriotic as he should be. He is one of the men who is responsible for present conditions. He should never growl if prices are cut in two in the middle. He must be aroused from his lethargy, indifference and inaction.

As a class, farmers are not hopeless. They can be reached and educated. This is the mission of the Farmers' Equity Union.

PEOPLE DRIVEN TO CO-OPERATION BY LARGE DIVIDENDS

"To recognize combination and monopoly as something necessarily here—square the law to the fact—and then, as a condition to granting corporate power at all, reserve the right to regulate dividends," is the remedy Judge Peter S. Grosscup, of Chicago, advances for Trusts. This is certainly a start in the right direction. The best thinkers in our land are declaring that the Sherman act, even as now interpreted, is an ineffective remedy for dealing with combinations in the business world. Big dividends on money invested robs both producers and consumers. The workers, the wealth-producers, are wronged more by big dividends on the capitalist's dollars than in any other way. Combination and co-operation are both just and wise if founded on righteous principles. But combination and co-operation which enable trusts to declare dividends from 30 to 40 per cent on dollars invested, is a robber system which should be checked by the government, just as high rates of interest by money sharks is prohibited by law.

Large dividends on dollars invested robs our farming districts of hundreds of millions of dollars annually, and centers it in the hands of millionaires, thus continually increasing their power for evil. The large dividend system is responsible for the centralization of the wealth produced annually in the hands of the few. Prevent large dividends on stock subscribed in all corporations, and we will have a more equal and just distribution of the wealth produced in our country.

When a corporation can declare a dividend of 50 or 80 per cent on dollars invested, it has robbed somebody, and the guilt is no less when the large per cent is made by a farmer's elevator company, than if made by the Standard Oil Company. I find many such elevator companies in the northwest with no stock for sale. One set of farmers is robbing the rest. If we can educate the people against the big dividend system they will overthrow one great curse of our land and set us one notch nearer the millennium.

Corporations should be neither destroyed nor be allowed license to follow their own sweet will. They should be checked, controlled, regulated by a federal incorporation act which would reduce their dividends to a reasonable basis. Why are laws necessary for the control of Trusts? Simply because they are founded on the wrong principle of the largest dividends possible on dollars invested. The almighty dollar is the unit pushed continually to the front by human greed. Greed crushes out of many millionaires all love for humanity. The love of money is the root of all evil. We are glad to see the great economic Trust question agitated and discussed as never before, and we believe it will be solved intelligently by the American people. The Farmers' Equity Union will be one of the great factors in the solution of this great question. We are teaching a system of combination and co-operation that will need no control, check nor regulation of the few to the sorrow of the many. Neither is founded on righteous principles, and is therefore sure to live and grow.

We are on the equal rights platform. On the principle of the greatest good to the greatest number. Equal rights and equal opportunity for every human being is our creed. In our code the man is the unit and not the dollar. The triumph of our cause means peace on earth and good will to men.

We are educating the people against a system that enables a wealthy corporation to hold the price down on thousands of women and children who milk cows for a living, and up on hundreds of thousands of women and children in the large cities who must have milk to live, and all this in order that the milk trust may declare large dividends on dollars invested.

We are inciting the people to rise in their might against a capitalistic system that enables a few greedy millinonaires to control prices to both producers and consumers whereby they are enabled to declare enormous dividends on gold invested. We are convinced that there are selfish millionaires made so by greed, who will indeed crucify humanity on the cross of gold if allowed power sufficient. That power to enslave the people must be broken. The combinations of the few must be met by the combinations of the many, which follow true co-operative principles.

We are organizing 100 farmers at each good town and educating them to be their own capitalists, and to be true co-operators. We organize them into a co-operative stock company, build an elevator, handle grain, cotton, stock and all they sell, also flour, feed, fencing, wagons and farm machinery on the same margins as other dealers. Out of the gross earnings we take expenses, repairs, dues and 5 per cent dividends on stock; and all over this is net earnings, which are prorated back to stockholders according to patronage. If all corporations were prohibited from declaring over 5 per cent dividends on stock on actual valuation, wealth-producers would come into their own and the number of millionaires and paupers in our fair land would be greatly reduced. Every farmers' elevator company ought to come down to this principle at once and allow every farmer to take stock. Equal rights to all and special privileges to none is our principle.

Farmers will unite on this principle because it is right. Buy and sell on a safe margin and prorate all earnings over 5 per cent to stockholders for patronage. This enables you to demonstrate clearly in dollars and cents to each patron, annually, the great advantage of the new over the old system of doing business. It shows the outsider the benefits and brings him in. It is important to have good management, a safe margin, a small dividend on stock subscribed, as large a sum as possible to prorate for patronage. These all conduce to success. They bring the volume of trade which reduces the cost of handling to the minimum. It knocks out the large dividends which rob the people and gives the individual worker all he produces. The plan is practical and just.

FEDERAL GOVERNMENT THE UMPIRE OF INDUSTRIAL CORPORATIONS.

Attorney General Wickersham, before the Minnesota State Bar Association, July 19, 1911, took an advanced stand on Federal regulation of corporations, and declared that a government commission to regulate great industrial organizations in the same way that the interstate commerce regulates railways, was certainly most desirable, and that it might be absolutely necessary. He said the law of supply and demand no longer controls prices in the United States. For years, he declared, the prices in all the great staple industries have been fixed by agreement between the principal producers and not by a normal play of free competition. With the weight of an administration officer behind them, his remarks are significant.

The fact that thirty million farmers and their families in our country are paying arbitarary prices on all they buy annually, has been apparent to thinking people for years. It also is just as evident that our present methods and systems by which farmers' products reach our central markets, prevent farmers from having any power whatever to price their valuable crops, so that the great economic problem before the American farmer is: "How can we defend our selves against low, unjust prices on eight or nine billion dollars' worth of valuable products annually produced on our farms, and at the same time against high extortionate prices on billions of dollars' worth of necessaries which we buy as consumers?"

The farmers are the largest class of wealth-producers and consumers. In their unorganized condition they suffer the loss of untold millions of dollars every year by selling their most valuable crops of finest quality below cost, and at the same time purchasing enormous supplies at arbitrary, unjust prices. These facts are obvious to all reading farmers themselves, and yet they must be continually thrust in every farmer's face till he will act on a proposition which brings relief. The great objection against the rural free delivery of mail was "It can't be done," but the Grangers agitated and educated until it came, and millions of farmers receive their mail daily as a result.

The same old "can't" phrase is continually raised against the Equity Union movement now slowly, but surely being planted in Indiana, Illinois, Missouri, Iowa, Kansas, Nebraska, South and North Dakota and Montana, but the present agitation will be continually increased as our means increase until one thousand successful Equity Exchanges are firmly established and demonstrating the practicability and justice of true golden rule co-operation in all these states. Then the sailing will be easy, progress rapid and success assured.

The steps are these, viz: Five farmers must unite and get a charter as a local union of the Farmers' Equity Union. They must get ten new members. The number must be doubled again and again. They must meet regularly the first Saturday of every month. One dollar annual dues must be paid. All of these "musts" must be obeyed. When one hundred members are enrolled an Equity Exchange must be organized. Every member ought to read the "Equity Text Book." He must be educated to be a Golden Rule Co-operator. When enough capital is paid in, elect a board of five directors and get a charter from the state. Build a good elevator with modern equipment. Buy and sell on a safe margin. Never declare over 5 per cent dividends on the stock subscribed. All over is net earnings, and must be prorated back to stockholders according to patronage. Count out in cash all you can annually to each stockholder for his patronage, so he will see the difference between the old and the new system. Count it out to him in cash so he must see it, and his neighbors on the outside will see it, too, and come in.

Allow no man to take shares unless he owns land or lives on a farm. Patrons must run this business. Shut out Mr. Capitalist. He wants to live without work. We can't afford to support him. We have too many like him now. We need more wealth-makers, but not any more wealth-takers. We produce plenty of capital and are hereafter going to control the capital we produce.

Allow no man to hold over four shares of $25 each. Allow no shareholder to draw out anything for patronage or interest until he has four shares—the limit. Pay him what he earns by his patronage, in shares, and increase your capital till 200 farmers have exactly $100 each in the Exchange, or a capital of $20,000. Then you have good financial standing in the business world, and if you have a fire you are able to build again. Never have a sinking fund or surplus fund. When you find it necessary to increase your capital for the purpose of branching out into farm machinery or lumber, as you will, increase the number of shares to five instead of four, and increase your capital out of your patronage. Keep the shareholders equal in vote and the shares at par value. Be sure to have a cut off January 1st and July 1st, and have the books examined by an expert and a full report made to the stockholders. This is very important. Remember, your manager is umpire in this game, and must be respected till you get a new one.

One thousand such Equity Exchanges all linked in the great Equity Chain will be potent demonstrations of the righteousness of the cause and practicability of the plan. They will speak more clearly and forcibly than all previous lectures, literature or books on the subject.

Then the entire output of mills, mines and factories can be purchased one year in advance to the great advantage of factories and farmers, and high extortionate prices will be at an end. Shipments of our crops to central markets will be regulated, gluts prevented and fair prices maintained for both poor and rich farmers alike. The farmers' business will then be the best in our country, because he has made it so himself through organization, education and co-operation. The farmers must put themselves on the same industrial level with the rest of the business world. We must be a great Industrial Union organized on the principle of the greatest good to the greatest number. All industrial classes must be organized for the promotion of the intelligence, morality and fraternalism of the members and persuade them to own co-operatively the controlling interest in the mines, factories and railroads.

Industrial Unions must co-operate for the welfare of all wealth-producers. Mr. Wealth-Taker must be put out of business. Industrial corporations are right and proper, and we are not afraid of their size, but they must be owned, controlled and run by the people who operate them on the principle of equal rights and opportunities to all. They must be controlled by the intelligent majority of wealth-makers and not by the few wealth-takers. Then they are in safe hands, and will be easily regulated by the Federal Government, made up of representatives of Industrial Unions as the umpire in the great game of corporations.

THOROUGH ORGANIZATION AND EDUCATION

One thing which ought to commend the Farmers' Equity Union is the thoroughness with which it strives to do its work of organization and education. Our organizers are expected to concentrate their efforts on a few places until one hundred are enrolled in each local union. Each member is expected to become an organizer until every farmer in his locality is brought into line. Every local union secretary is required to have two lists including all the farmers who market at his town, one a list of members and the other of nonmembers. Attention must be given at every meeting of the local union to stray or delinquent members, and committees must be constantly at work to enroll new members. Then it is the province of this union to carry on a continual campaign of education and encouragement for all of its members.

Sec. 2 of Article 2 says: "The local union shall be a regular farmers' club, promoting the intelligence, morality and every social interest of the farmer and his family."

Sec. 4, Article 2 says: "Every member joining shall be induced to become a subscriber to the Equity Union Exchange, our official promoter and medium of exchange. It is very important that our paper reaches every Equity family weekly for the instruction, information and encouragement of our members. The cost to each members is only a trifle. The paper is the key to the whole situation. If every member would read the Equity Union Exchange carefully, we would soon work out and carry out a plan of co-operation worth millions of dollars to the farmers.

Every local union must have a correspondent, reporting all progress, encouraging news, crop conditions and prices. Let us fill the Equity Union Exchange full every week with helpful reports and suggestions, and of new locals formed, new members received, successful meetings, picnics, rallies, declamation contests, corn shows, and especially new elevators built. Every member must be induced to subscribe for the paper.

The Equity Text Book, "Farmers Must Be Co-Operators," ought to be in every farmer's home. It educates. It gives a full complete explanation of the Equity plan of co-operation. It contains the constitution of the Equity Union and By-Laws for an Equity Exchange. It is a complete guide to true co-operation. A regular bureau of lecturers on conservation of soil, of patronage, of political power, on corn culture, dairying, improvement of country schools and beautifying of country homes, must be organized and kept constantly in the field.

The educational feature must not be neglected nor overlooked. Economic freedom is only possible to an intelligent, moral, fraternal people. Grafters, extortioners and speculators are banking on the assertion and hope that farmers cannot be united, will not stick, and never will be co-operators. But we have faith in the good hard sense of the American farmers. We believe their suspicion, prejudice and ignorance can all be overcome by a continual campaign of education. The process must necessarily be slow, but it is sure and brings permanent success and lasting benefits.

Our purpose is to have a Farmers' Institute at every good town once every month, in which farmers and their families and educators and lecturers unite in the discussion of all subjects of special interest to country people. This will bring a general uplift by farmers and for farmers. The three M's are absolutely necessary to the success of this grand movement, viz: members, meetings and money. These must be provided.

It will pay to quit work the first Saturday of each month and go to the Equity meeting. "Farmers are too busy." The busy farmers who have no time for an Equity meeting fatten the trusts, support the speculators and make hunderds of millionaires each year.

A picture represented the busy farmers cultivating a big apple tree bearing fine, large, red apples. But the farmers were so busy plowing, harrowing and cultivating, that they could not be induced to look up in the tree and see the big fellows with plug hats gathering all their apples. On the side of the tree was a ladder, "The Equity Union," reaching from the ground right up into the tree, and the union farmers were climbing the ladder with clubs in their hands to knock out the robbers in the tree. We cannot afford to neglect our Equity meetings. Thoroughness of organization and education of the farmers at every good town will result in strong Equity links that will demonstrate the power and benefits of true co-operation and unite the farmers against every foe and make them invincible over every enemy.

BIRD CITY MEMBERS LOYAL

We want every member to read the letter by Brother Henry from St. Francis, Kansas, in this book. "The enemy" tried to buy our members at Bird City, Kansas, for seven cents a bushel and failed. A Union made of that kind of material will succeed without any question. Bird City Union has "a sticky bunch." They, however, deserve special mention when they are tried in the balance and not found wanting. Seven cents a bushel is a big bribe but that grain company has run against a different kind of Union from any they have ever seen before.

Our members, especially Kansas members, are educated co-operators. The Equity Union has only enrolled about ten thousand regular members, but those who have read our co-operative paper weekly and attended our meetings are educated to attend and stick to their own business. We know what we are trying to do and we stay with it. Equity Union members understand clearly that there is a great difference between our system of business and the old graft-profit-system which robbed them of thousands of dollars and they can not be bribed with even seven cents a bushel to support the old system.

We say, hurrah for the Bird City Equity Union. H. Z. Baker is their president.

SEVEN CENTS A BUSHEL WON'T BUY THEM

St. Francis, Kans., Aug. 29, 1914.

Equity Union Exchange,
    Greenville, Ill.

Dear Sir: We are just getting opened up in good shape. We need our national president to stiffen the backs of some of our stockholders. They can't understand why we can't pay as much as the old line companies while they are fighting us. We can't help ad- cents per bushel more than the exchange elevator was paying. Hurrah, for the Bird City Equity Union.

My idea was when we were offered more than a thing is worth, let the enemy have it. But I have been watching the workings around here in the different unions, and am convinced that if our members take a safe margin and keep away from the enemy entirely, we will do much better than to play with them, strengthen our organization and make more profit in the long run.

We are not commencing on every thing the first year. We must be educated and that takes time. When may we expect you here to speak to us again?

Yours truly,

C. R. HENRY.

EQUITY UNION MEETING, BIRD CITY, KANSAS

This is a picture of the Equity Union members at Bird City, Kans. They have over five hundred members in this Union as every member of each of the one hundred families is a worker for our grand cause. There are whole families in this group, who would do without 10 meals each month, if necessary, to support the Equity Union. If you want to find men, women and children who are loyal to our great cause, go to Bird City. They will own their grain elevator, warehouse and coal sheds and buy and sell together. No Line Grain Company need apply for Equity Union grain in 1914. They refused a bribe of seven cents a bushel of wheat in August 1914.

OFFICERS AND BOARD OF DIRECTORS OF MOTT, NORTH DAKOTA EQUITY UNION

The Mott Equity Union has made the best showing of any Equity Exchange in the country. They were one of the first Unions organized and were a little alarmed when they found their charter was only number nine. But a strong majority of the membership had the intelligence to see that the Equity Union was established on just principles and advocated a practical plan of Golden Rule co-operation which was bound to win if pushed by loyal members. Our weekly paper was read by a majority of the Mott members, and showed them that there was a National Union behind the movement pushing Equity Union into twelve states and establishing Exchanges with 100 or more good farmers in each one who would stick together on this plan. Our National Union is an organizing, educating force. The Mott directors deserve especial credit for the erection of one of the finest elevators in the state and for handling the 1912 crop of grain so as to save the members over nine thousand dollars in one year. Give us 1000 Equity Unions like Mott, N. D., and we will demonstrate results that will bring in one million members in a short time.

A LITTLE HISTORY FOR NEW MEMBERS

Three plow handle farmers from Illinois started the Farmers' Equity Union December 16th, 1910, by getting a national charter from the Secretary of State, at Springfield, Ill. Under this charter we can charter a local union anywhere in the United States.

Read the By-laws in this book. They were revised by the national meeting in Chicago in December 1912, and again by the national meeting at Kansas City, Kansas in 1913. They are subject to change at any national meeting by regularly appointed delegates from the Local Unions. The members make the laws which govern them.

This Union has members in fourteen states. If you read the By-laws carefully, you will see that we have a well defined, practical plan of co-operation. We center our forces, labor, efforts, money, and time on the education of each community, where we start a Union, so that locality will carry out fully our plan of golden rule co-operation in business. We try hard to organize each Local Union into a stock company with 100 or more stockholders. The shares are $25 each, no member can hold over four shares of the stock, and neither can a member draw out any of the earnings of the company until he has four shares. One hundred shareholders means a capital of ten thousand dollars. This money is invested in a grain elevator, warehouse and coal shed and some money in the bank for capital. A board of five directors direct the business. They hire a manager. He buys cream, eggs, poultry, grain. He sells flour, feed, coal, twine, salt, cement, etc., at the market price. He never boosts the price on what he buys nor cuts the price on what he sells.

He buys and sells at the SYSTEM PRICE, which gives him a safe margin. We work for the trade, not only of our members, but of outsiders, paying outsiders the same price as members. Every one of our Equity Exchanges which handled the 1912 crop of grain, had a fine gross earnings. Out of the gross earnings, the directors paid expenses, interest on capital borrowed, one dollar national dues for each stockholder and a 5 per cent stock dividend. All over this was net earnings or profit, which should not have been taken from the farmers, and was therefore paid back to stockholders according to patronage. The profit is figured as a per cent of the business furnished by the stockholders, and paid back as a patronage dividend. We run a SAFE BUSINESS, so that instead of being behind one or two thousand dollars at the end of the year, we were from one to nine thousand dollars ahead at each Exchange after paying all expenses and a 5 per cent stock dividend. This was our experience in handling the 1912 crop. The reports will soon be coming for the 1913 crop as we are nearing the close of the grain season.

MR. AND MRS. EDWIN W. REED, LUX, NEBR.

NATIONAL DIRECTOR FROM NEBRASKA

Mr. Reed is a stockman in western Nebraska, and is a member of Haigler Nebraska Equity Union. He is deeply interested in the Equity Union and will make a good member of the national board of Directors. We welcome him as one of our national officers.

We are glad to enroll Nebraska among the Equity Union states. We do not organize by states or counties but by railroad lines, but every state is represented on the national board of directors and has a voice in the control, management and direction of the entire National Union.

We believe we will find in Edwin W. Reed, a faithful, conscientious worker and amiable counselor.

THE WORLD DOES MOVE

An emergency has arisen which will make our government a common carrier. We have, as a nation, refused to subsidize the building of merchant ships for the profit of their private owners, but our President recommends the appropriation of twenty-five million dollars by the national congress for the purchase of merchant ships.

The government, as owner of the Panama railroad company, is already a common carrier both on land and sea, because the Panama road owns several merchant vessels. The present need of our commerce is imperative for ships to carry our surplus crops to Europe. There is a difference between subsidizing of ships privately owned, for the benefit of their private owners, and the government ownership of ships for the profit of the American people.

We have finally **moved up** to the high standard where the president of the United States is inherently and openly opposed to the policy of taxing all the people for the profit of the few, and we believe the people of our country stand with the President in this policy, and we believe the people are being slowly but surely educated to stand together against "special privileges for the few."

"Uucle Sam" is carrying our mail and our packages at **actual cost**, without profit, and will finally carry our grain, stock, cotton and all produce and merchandise, both on land and sea, without profit. **The world does move."** The people are learning how to eliminate graft, profit, high rates of interest and all special privileges. Equal rights for all and special privileges for none, will come as we **push Equity Union.**

It is a great thing to have a governmental system that is not cast in a hard and fast world of bureaucratic paternalism or monarchial absolutism. J. P. Morgan & Company is radically opposed to the whole proposition. But he and his kind have ruled the industrial world entirely too long in our country. It is time " the people" were awakened and aroused against the Money Kings. Equity Union is educating the people to run their own business. Our pass word is "Mind your own business." Get your neighbor to join the Union. You are working for "God and home and Native Land" when you work for this grand organization.

MRS. MARTHA REED CONNER, CEDAR BLUFFS, KANSAS

Mrs. Martha Reed Conner, Cedar Bluffs, Kansas is a regular member of the Farmers' Equity Union, owns a farm and is seventy-seven years young. She believes strongly in the principles of the Equity Union and has strong faith in the triumph of our great cause. Read her letter in this book.

MRS. CONNER'S LETTER.

Mr. Editor: At your request I am sending you my photograph and a short sketch of my life. I was born at Piqua County, Piqua Plains, Ohio in 1837. Was educated at Curcusville, Ohio. Married to John Conner in Baska Bell, Wis., in 1857. Moved to Dodge County, Nebr., 1868. Moving on to the western part to Red Willow County, Nebr., in 1879, and living close to the Kansas line, Cedar Bluffs, Kans., has been my post office. We have owned hundreds of acres of land and farmed on a large scale, and owned thousands of head of sheep and a great amount of other stock, and have had a good chance to know how the farmers are treated and I think from what I have studied about the Farmers' Equity Union that it will prove one of the grandest things for the farmer that was ever known.

Although I am getting up in years and at this writing have a broken arm, I am anxious to live and do what I can to see farmers come to the front, and believe the day will come soon that they will be the leading people instead of the hayseeds.

Yours truly,
MARTHA REED CONNER.

HOW MUCH WILL YOU GIVE—HOW MUCH WILL YOU TAKE?

When we think of the wonderful power one million United farmers would have in the business world and in the political arena, it is certainly pitiable and disgusting to think that the millions of so-called independent farmers are today standing in the markets crying: "How much are you going to give me, and what will you take?" Everything he sells and everything he buys is priced for him by a combination.

He has been educated to believe that the farmer is an ox or a brother to the ox and all he needs to know is how to work and the road to market and when he gets there a "Business Man" must weigh, grade and price all he sells and all he buys.

No wonder we find markets all over the land organized against the farmer, both in the country and in the central markets. No wonder that there are thousands of country and central markets which pay as little as possible for farm produce. As long as millionaire meat packers own the packing plants and central stock yards, the production of hogs and cattle will be discouraged, because the packers are not in it for their health. The country markets will pay as little as possible for fat hogs and cattle as long as they are run by profit-takers.

The farmers can and will produce a full supply of meat products when they market and pack their own hogs and cattle, and consumers' prices will be more reasonable. Stockmen must unite in the Equity Union and begin to ship all their stock in one channel. They must be co-operators.

Dairymen.

The milkmen are up against a strong combination. They pour millions of dollars into the laps of millionaires every year. That "Independent Farmer," in his overalls, check shirt and slouch hat, meets Mr. Millionaire Profit-Taker and meekly asks: "How much do you think you can give us for milk the next six months?" Mr. Profit Taker says: "We'll pay you as much as the rest." But how much is that? As little as we **dare**. Will that price pay losses, labor, feed and a little each year for old age? That side of the question is not considered one moment because the farmers have nothing to say in pricing the milk. The question is "How long do we **dare** to hold the price on these industrial slaves?"

The object of the few millionaires who run the milk business for us is **big profits and swollen fortunes.** Their plan is to hold the price down on milk producers and up on milk consumers. The individual farmer has no power against such a combination. His only hope is in the Farmers' Equity Union.

If the milk producers will unite in the Equity Union we will receive a just price for milk, that will cover losses, labor, feed, expenses and a little profit for old age, and at the same time give consumers more reasonable prices. This is done by dividing the enormous profits now taken by the millionaire profit-takers in the milk business for all there is in it. The meat trust and milk trust will have power as long as farmers refuse or **neglect** to unite. Their power is gone the very day that the farmers unite.

Give me a solid union of farmers in Bond County, Ill., and milk goes up one cent a quart, 4 cents a gallon, 50 cents a hundred or one hundred thousand dollars a year to the farmers and down the same amout to consumers.

Farmers, listen! We are to blame for bad conditions on the farm. We have it in our power to change the conditions whenever enough unite. Wherever the Equity Union is strong the farmers do not say, "How much will you give nor how much will you take?" They have a market that pays as much as possible for all produce and sells as cheaply as possible. No profit is taken when you buy or sell. Mr. Profit-Taker must hunt another job. Mr. Millionaire has enough and is out of business. The wealth produced goes to the people who produce it. We quit making millionaires. We are no longer their industrial slaves. It is much better to unite and become brothers.

Equity Union will make us free men. It will build hard roads, good schools and churches and comfortable homes and well improved farms in the country. Not so many mansions and autos in the cities but more in the country.

KNOCK DOWN THE TOLL-GATES

The great economic question before the American people is the distribution of wealth. Pat said the rich ought to be forced to divide with the poor. Some one said, "Pat, that saloon would get your share in about six months and what would you do then." Oh! I'd divide again," said Pat. This is one idea of the distribution of wealth, but not the intelligent thought of the American people.

The Equity Union believes that every man ought to have all that he producers, and his children ought to go to school and his wife ought to make his home the best and happiest place on this earth. This is ideal and will come when enough golden rule co-operation is instilled into the minds and hearts of the people, so they will not only pray and preach it on Sunday, but live and practice it in business, six days in each week. This is practical Christianity that will bring the Millennium. There is great cause for the discussion and solution of this question of the right and just distribution of wealth in this, the richest country in the world.

We do not believe in the anarchist hobo, the I-Won't-Works or the I-Want-Whiskeys. But there is a more dangerous class to the economic freedom than the above classes. The first named classes can be put on the rock piles and forced to earn their bread. The police, and if necessary, the state militia can handle these classes.

But there is another rich, powerful class in the business, social and political world, who rob the wealth producers of billions of dollars annually in this land of the free and home of the brave. We do not believe in condeming well-to-do people by the wholesale. We believe there are millions of men who have amassed a fine competence for their old age by honest industry and thrift.

But the rich, powerful class to which we allude are entirely a different set of business people. They are after a great fortune at any cost. The old man said, "Boys make money! Make it honest if you can, but boys **make money!**" This class of men have organized and combined and co-operate for the express purpose of taking **billions of dollars** from the honest industrious millions of wealth producers of our nation.

Today the rich millionaires rule and influence the White House and make laws at the capitol; courts of justice are their ministers; senators and legislators are their lackeys. They control preachers in their lecturers and editors in their sanctums. They swagger in the drawing rooms and social circles and rule at the clubs and lodges. They dominate with a rod of iron the business world and the affairs of society.

Every year seems to enlarge their power, and the men and women who protest against the crimes that are being committed by organized greed in this country—who talk of protecting the American people—are ostracised and vilified—are hounded and imprisoned.

This class of organized capitalists has an insatiable greed for gold, which knows no bounds. Their love for gold drives out of their bosoms all love for humanity. They get control of the necessities of life and establish toll gates between producers and consumers and in a few years 90 per cent of them will be home-owners and home-builders and make the country the beautiful and ideal place it should be for people to live in.

Denmark has made home owners through golden rule co-operation. The salvation of the American farmer lies in a national union that educates up to the same grand principle. This slow but sure process of education is our only hope, but farmers are getting more of it than ever before, and along the right line, at last.

We must unite and do for ourselves the things we cannot do single handed and that we need NEVER expect others to do for us. God helps those who help themselves.

Let us advocate banks "by the people and for the people." Then buy together at jobber's prices for cash. Farmers, it CAN and WILL be done. Do not wear out the hold back straps. Come in and help build a great National Union that will secure banks that will loan us money at a low rate and then discard the credit system as far as possible.

Capital is as necessary in farming as in manufacturing and should be furnished as cheaply. But good conditions must be made on the farms by the farmers themselves. The Equity Union offers the plan of co-operation that will unite 1,000,000 farmers.

TWENTY EQUITY UNION FORTRESSES

Here is a good sample of the twenty fortresses the Equity Union is erecting on the Rock Island railroad in Southwest Kansas. Each fort is manned by six or seven hundred men, women and children, all Equity Union members. Our Kansas children are enlisting and are training for the Equity Union army.

They are reading the "Equity Union Exchange" and having instilled into their minds and hearts the principles of true blue golden rule co-operation. They will be our future great army, fighting for economic freedom.

These twenty fortresses are for the protection of the Equity Union families who must market about seven million bushels of grain each year when their crop is good. The Profit-takers have erected fifty or sixty fortresses of their own, at these twenty stations for the purpose of TAKING from three to five hundred thousand dollars PROFIT out of each crop and carrying it away to their homes so their families can revel in luxury and ease. Most of the families they ROB are poor and struggling for homes of their own.

Our twenty forts will be sufficient to cope with all the enemy's forts if we can get the people EDUCATED thoroughly in the principles of the Equity Union. This campaign of organization and education must go on incessantly, continually winter and summer. The PEOPLE must be gathered together for dinners, for sociability and fraternalism. They must be aroused, encouraged and instructed. THEY MUST HAVE FAITH IN EQUITY UNION. We must put more enthusiastic, able speakers in the field.

and force these two great classes to pay tribute to them as their masters or kings.

The meat packers force prices down on producers and up on consumers. The milk trust fixes the price for the women and children who milk cows and for the women and children who must have milk.

The flour trust makes a difference of twenty to forty cents a bushel on wheat, between producers and consumers. Producers at this writing are receiving 16 cents a dozen for eggs in the country, and consumers in the city are paying 32 cents a dozen.

Our only hope is a great industrial Union like the Equity Union which changes the entire business system from capitalistic to co-operative.

UNITED STATES BANKS VS. THE CREDIT SYSTEM

A Pettis County, Missouri, banker is reported as saying in a speech before the bankers at the Boston convention the following about farmers:

"Only a few years ago farmers secured the virgin soil at a low price, but today they are buying a depleted soil at a high price. It takes from four to six times as much to equip a farm for successful operation at the present time as it did 40 years ago.

"There is not one farm in fifty that has sufficient working capital to make it as profitable as it should be, and conserve the fertility of the soil at the same time.

"The conditions in this country at the present time is that mortgage indebtedness on the farms is on the increase, and under the plans by which these loans are arranged the payments NEVER WILL BE MET BY THE RETURNS FROM THE LAND.

"Production per capita is on the decrease, land tenancy is on the increase. In Denmark 89 farmers out of every 100 own the land they till, but in this country our land was almost given us a few years ago."

This level headed business man has given us a few facts worthy of serious consideration.

The farmers must have working capital sufficient to farm right.

A million farmers united in the Equity Union would elect a Congress that would put a United States bank in every good post office, that would lend money on good security at 3 per cent. No run would ever be made on these banks. Our deposits would be perfectly safe in them, and farmers who are honest and industrious could have capital enough to pay cash for what they bought. We would then buy together in large job lots direct from mills, mines and factories, at about one-half of present prices. Millions of dollars would stay in farmers' pockets which now go to support the present outrageously expensive and extravagant business system.

The Equity Union system is the Denmark idea of co-operation.

Give us a million co-operating farmers united in Equity Union

The twenty-five Equity Union rallies held on the Rock Island railroad in April and May have been a wonderful power for good. Great enthusiasm has been engendered and a determination is aroused along this section of Kansas to overthrow the enemy's forts and establish firmly our own. If the enemy gets a bushel of Equity Union grain this year, it will be because we cannot handle it ourselves. This is the feeling in this "neck of the woods."

Most of our members are poor but they are honest. The enemy cannot bribe them with a few cents a bushel. They will be loyal to Equity Union. We do not believe there will be a traitor in the camp. They will stand firmly by the Equity Union army forts. We are going to extend our line of forts clear into Kansas City on the Rock Island and then out on the Sante Fe and all the railroads.

Brothers, go after those farmers on the outside. They must not be left to support the enemy. They belong to our army and must be enlisted. Let the Equity Union army of the United States cheer up and go after more recruits. There are thousands of good farmers who can be enlisted in this fight for economic freedom.

Do not be afraid of the enemy. They have no weapons but ABUSE and BRIBERY, and these will not win with the honest intelligent people, we are enlisting in our Equity Union army. Be of good cheer! The victory is to the brave and the true. Kansas expects North Dakota to stand with her and have as many Equity Union fortresses as she has.

SUPPORTING A UNION

The farmers are supporting all kinds of Unions at heavy expense. These Unions work right against the farmers and take millions of dollars every year from them and yet the farmers support them.

I find the farmers supporting a Union of grain elevator men at hundreds, yes thousands of good grain markets and paying thousands of dollars for their support. I know they are supporting a Union of millers and paying more than one hundred million dollars for it annually. Every reading thinking man knows they support a Union of meat packers to the tune of 100 million dollars annually.

I know of nine large communities of farmers in Illionis, Ohio, Wisconsin, Pennsylvania, Kansas and Colorado, who support a little Union of about two dozen milk condensers and pay a half million dollars annually for the support of that little capitalistic Union?

Of course these nine bunches of "independent?" farmers are only a small per cent of the number all over our country who are doing the same thing. How many million dollars do we pay annually to support the great army of traveling men's Union! What is the cost to the farmers for the machine trust union, the coffee trust union, the match trust union, the salt union, the clothing union, the lumber union and all the other unions for somebody else, which our poor deluded unorganized farmers pay? But listen, brother farmer!

When they are asked to pay one dollar per year for a union for themselves they absolutely refuse.

They seem to prefer to support everybody else's Union rather than their own. One big community of rich farmers in Kansas actually adopted our by-laws to try to unite the farmers locally, but refused to come into our National Union and support it with one dollar per year.

They are penny wise and pound foolish. When enough farmers are wise enough to support the Equity Union with one dollar a year, we will stop draining the country districts of hundreds of millions of dollars which go to support other people's powerful unions, which grow stronger every year by the support of millions of foolish farmers.

The farmers who will support the millionaire union with one hundred dollars per year and refuses to support a farmers' union with one dollar per year, brands himself as either a pin-head or a hay-seed. The Equity Union makes the three dollars entrance fee right back for each member on twine, coal or something else and makes the annual dues for the member in the Exchange.

The fees and dues are all made back over and over, again and again and yet we find it very difficult to get all of the members to be prompt in paying their dues.

The national president has worked hard for seven months of 1914 without drawing a cent of salary. The expense for organizers, for headquarters, and for starting the new paper have been heavy this year. If every farmer who has joined the Union would pay his 1914 dues we would have enough money to push the work much faster.

We want to thank the Equity Exchanges that have paid-in the national dues for every member. This is the only sure way to get support from the farmers for a National Union. Our By-Laws provide for this and the provision is a wise one.

It will help us to build a strong National Union of intelligent progressive farmers that will break the power of every Union of profit takers, grafters and extortioners.

We will have to leave the pin heads and hayseeds on the outside to support other peoples' Unions, till they learn better. Some people only learn by hard knocks. You little narrow suspicious fellow, just stay on the outside and take your medicine. Every Equity Exchange must see to it that only members of the Union are allowed to be stockholders, and that outsiders are allowed no benefits from our Union whatever. This is the only way to protect and build a successful Union.

A CONVERSATION BETWEEN TWO OLD LINE ELEVATOR MEN

Bill: Bob, how do you get along with the farmers' elevator companies? Can you compete with them?

Bob: Not with those Equity Union companies. They stick together, buy on a safe margin and pay back all profits. How are you going to compete with that system?

Bill: Why, just bid above the market a few cents for one year and they will all let their elevators stand and give us their business.

Bob: There is where you are mistaken. We bid 3 cents a bushel above the market last year and their 170 members at that point went straight to their Equity Exchange with every load of wheat.

Bill: Well Bob, this is getting serious. They are going to let us down and out. We must get them to wrangling among themselves. This is our only hope.

Bob: Yes, if you can. They were in a little split at New England but their national president held "a get together" meeting there July 9th, and their State Organizer another one July 18th, and New England Union is going to be one of their strongest Unions. They are going to get together.

Bill: Yes, that National Union plays the dickens with our business. They have worked out here in Dakota for four years and put a rank socialist paper in the homes of all their members. How will we fight that Union?

Bob: The only way I know is to show the farmers that the organizer is just working them for the money there is in it. Some of these farmers are awful suspicious.

Bill: Bob, that game is playing out with these Equity Union farmers. They read and think for themselves. We fooled all of them for awhile but since they went in this new Equity Union we can't fool only a small part of them. You can't buy either them nor their grain.

Bob: Well we have 100 elevators and seven big mills and millions of dollars of capital and we will kill this enemy of our business if we have to spend a million dollars to do it.

Bill: Bob, there is where you will run your head against a stone wall. I have investigated this new organization. It is a young giant. They are not very large in number but in principles and a practical business plan. They are ahead of anything we have ever had among the farmers and their Union is doubling in membership every year.

Bob: Yes, but we have the capital, the elevators and the mills.

Bill: And they have the **grain** and will have the mills too.

Bob: But we will go right into their market and undersell them with our flour and feed.

Bill: Where will you get the wheat to make your flour and feed?

Bob: We will buy it from the "scab" farmers.

Bill: Well, Bob, this Equity Union is a strong organizing force that will organize a market among its own members in the country and in the cities that will not eat your trust ridden flour made from scab wheat. The Equity Union has applied for a copy right from Washington on "Equity Union Line" and will brand it in every one of their sacks so their members will know their own flour. You can't compete with an organized market.

Bob: I know the people will buy our flour if they find it is 50 cents a hundred cheaper than the Equity Union brand.

Bill: That was true generally at one time, but that day is passing. The Equity Union is a growing educating force. It lifts the people up in intelligence and morality and fraternalism until the trust can't bribe any number of them to stand against the interests of the entire community.

Bob: Well they will have a good time taking our business away from us.

Bill: Bob, "the people" are aroused on this trust business, which has robbed them so long, and are organizing against it into industrial Unions, which will break the power of every trust in America. I am a mill-wright by trade. I am going to work for "the people" after this and you better get into the Band Wagon, too.

PERSONAL LETTER TO THE KANSAS MEMBERS

The picture shows a familiar scene in Southwest Kansas. It is Kansas. The sale at a fair price of this magnificent crop of golden grain is the question in which the Equity Union has a special interest.

If the Kansas farmers market one hundred million bushels of wheat in 1914 through the old profit-taking system, the grain companies will TAKE from three to five million dollars unnecessary profits. Kansas farmers will be short millions of dollars while a few rich grain companies will be made stronger than ever to rob the peo- on the farm of G. A. King of the Naron Equity Union in Pratt county ple. This is a big robbery of a class of people who work for every dollar they get. It is a shame that our progressive Kansas farmers have not been shown a better method of marketing long ago.

Millions of dollars have been TAKEN out of their big crops by millionaires who sow not, neither do they spin, and yet Solomon in all his glory was not arrayed like one of these. The Kansas farmer is not ashamed of his check shirt and overalls but he is tired of putting broad cloth on the backs of a few parasites.

If the Equity Union was strong enough in Kansas to market the entire 1914 crop of wheat on our plan of golden rule co-operation, we would prevent this enormous steal and the farmers of Kansas would conserve their prosperity to the extent of millions of dollars.

A great injustice would be prevented and a higher civilization in the country would be the legitimate result. Grand country highways, beautiful country homes and modern rural churches and central district and high schools will all come as a result of golden rule co-operation.

If the fifty Equity Unions started in the great grain state of Kansas are built up solidly by the hearty co-operation of the National and Local Unions and Equity Exchanges, and by the faithful work of all officers and members, we will soon have a demonstration of golden rule co-operation at fifty good markets, in that state which will unite and keep united more than twenty-five thousand men, women and children in an industrial union that will be as a city set on a hill that cannot be hid.

The Equity Union worked hard for practical demonstration of our plan and principles at one good Kansas market in 1912, and succeeded to the extent of a seven thousand dollar patronage dividend and a membership of about 150 loyal regular members.

This successful demonstration at one good market in Kansas has given us fifty new Unions in that state which are it least started on the Equity road to success. Now the building up solidly of these fifty Unions and the starting of fifty more in the next twelve months in the great grain state of Kansas is the purpose and full determination of the irrepressible Equity Union.

We ask and entreat for the heartiest and fullest co-operation of every Equity Union man. woman and child in the Sunflower state. A full and complete demonstration of Equity Union golden rule co-operation at 100 good markets in the progressive state of Kansas will be an educator that will redeem the state from the graft business system that now robs its best class of citizens of untold millions of dollars annually. We are writing this personal letter to our thousands of Kansas Equity friends as we speed along on the Burlington train on our way to the Dakotas to do what we can to build up the fifty unions up there during June and July.

They are all grain growers and have a common interest with the farmers of Nebraska, Colorado, Kansas, Oklahoma and other Equity Union states. They need Equity Union co-operation and are beginning to go after it in a very practical way.

ORGANIZING UNDER ONE NATIONAL HEAD

There are said to be two thousand farmers' elevator companies in the United States. More than seventy per cent of them are only organized locally. A large per cent are capitalistic. There is very little genuine true-blue co-operation among them. There is so little of the co-operative spirit in the minds and hearts of the members that it would be impossible to unite them under one national head on a true blue co-operative platform. These so called companies cannot survive the storm of competition that is coming from the united opposition to all farmers' organization. They do not unite "the people." Their principles are steeped in selfishness.

There is a per cent of the farmers' companies, however, that are truly co-operative. They only declare a nominal stock dividend, not over 5 per cent and prorate all of their net earnings as a patronage dividend and they are uniting the farmers strongly in each locality, but only locally. Now the Equity Union is not only uniting a large per cent of the farmers in each locality, but we are uniting them under one National Head. This is the great need of the farmers of this country. It is well to be strong locally, but not sufficient. The great trusts are organized nationally and can only be met squarely by a National Union of farmers and wealth producers.

Give us 200 Equity Exchanges with 100 farmers in each one and all firmly and loyally united under one National Head and all buying all of their farm machinery together, and the benefits would be so marked that twenty thousand more will tumble over each other to get in. Then forty thousand farmers will buy together and the results will be so big and so apparent that even the "hayseeds" will sit up and take notice and forty thousand more will come marching into the Equity Union, and every big machine company will be chasing after the Equity Union business, because it is organized nationally instead of locally. Slowly but surely we are getting our reading, thinking members to see this. That we must organize nationally under one head and only one head. We must have Local Unions and Exchanges all united under one head. Have no county, district nor state unions to support or to creat jealousies, friction and division. These Exchanges must buy and sell together as much as possible. The more our patronage is organized the better for each individual member.

A strong **national head** composed of representatives from each large strong Equity Exchange is the organizing, educating, directing force we need and that we are getting in this great organization. Every state is represented on the national board of directors, and every Local Union or Exchange is represented in the National meetings and has one vote for every live member in his Local Union, so that it is a peoples' organization united under one national head. We have no state, district nor county Unions to support or to create friction or division. There is only one great directing head in Equity Union and it is under complete control of the members so far as they choose to act. This fact is bound to make the Equity Union a powerful organization. It is **the growing giant** with which grafters, extortioners and profit-takers must reckon in the future. The "money changers" in the holy temple are being driven out already and they are cursing our new system which is for the people. Our Union is simple and inexpensive but powerful and far reaching in its possibilities for the common people. The A. S. of E. collects five dollars entrance fee and five dollars annual dues. Our entrance fee is only three dollars and our annual dues, one dollar. We have no state, district nor county officers absorbing our revenue and creating friction in the working field. There is no place in the Equity Union for the "hangers on" to cling, or to get a hold. Too bad, isn't it!

Let us keep our Union clean and simple. All of our Exchanges must pay the national dues for each member, as provided in the by-laws. This method is simple, inexpensive and sure. Let us unite strongly under one National Head. **Never take a man as a stockholder in an Equity Exchange unless he is a member of the Union. He has no business in our Exchange unless he is a member of the Union in good standing.**

DEFINITE INDIVIDUAL RESPONSIBILITY

The success of the Farmers' Equity Union depends upon the success of our Equity Exchanges. Every member of the Union who is a farmer ought to be a stockholder in the Equity Exchange. Stockholders should meet often and hear reports of the way your business is run. You are responsible for seeing that the business is run honestly and efficiently. Great private business enterprises owe their efficiency and success very largely to the fact that they are so organized so as to create definite individual responsibility for the accomplishment of their purposes. This definite individual responsibility is a feature which must not be lacking in the government of our Equity Exchanges. The sovereign power and responsibility rests with the stockholders. If they see that a director is not in sympathy with true golden rule co-operation or is prejudiced against the union, the sooner they weed him out the better it will be for the Exchange.

Our success depends on a very large volume of trade well managed. Work for honest efficient management and a very large patronage. Bond the manager, balance his books weekly. Buy and sell on a safe margin. Do not cut prices. Sell flour, feed, fertilizer, fencing, wagons and farm machinery at the same price as the other dealers. This insures honest management and a safe business. Hire a good manager and pay him a good salary. He must be a good business man and a good mixer. He must know how to handle grain and how to handle men. This will give successful management.

Prorate back all profit on the business to stockholders according to their patronage. This will down the profit-system and overthrow the capitalist. It will unite the farmers more and more. It will bring more members, more stockholders and a large volume of trade. Never prorate to the outsider. His influence is against us. We must bring him over to our side. Hold the profit system to which he clings so tenaciously over him, until his co-operating neighbors jingle the coin before him year after year, which they have gained by co-operation, and he will finally come into our camp.

Nothing will tempt the manager more than to see that his directors do not direct. The directors will be tempted to go wrong when they see the stockholders are careless or indifferent about the business.

We are not aiming to insinuate that Equity Exchange stockholders are careless, nor directors or managers dishonest, but we are very anxious to see every Exchange in the United States carry out the principles of golden rule co-operation and be a grand success. A large volume of trade and honest efficient management are a guarantee of success every time. We must work for these. There is no conflict in our country between labor and capital, none whatever. The uprising in the industrial world is not against capital but against the capitalists who have combined to make industrial slaves of the laborers. The success of our Equity Exchanges will mean the complete overthrow of the capitalistic system in the business world. Every true Equity Exchange will have plenty of capital but not a single capitalist.

The golden rule spirit is that every man shall have all the wealth he produces. The Equity Exchange protects him against the capitalist who would take part of what he earns. Farmers! We have robbed ourselves and our families long enough by supporting the present business system. We have paid for automobiles, fine homes and luxuries for the other fellow and his family while our families have had only a bare existence. It is our fault. Our separation is our weakness. We must unite and co-operate. We must see that our Exchanges carry out the Equity Union by-laws. If those directors do not follow our by-laws, throw them down and out. We, the people, are sovereign. Those we elect to represent us must not be our masters but our servants. Let us meet once a month, shake hands, become acquainted, be more friendly and more fraternal and so organize our Equity Exchanges as to create definite individual responsibility on the part of every director, every officer and every manager.

INDEPENDENCE DAY

This day has been coming to the United States for 138 years. It is always a day of gladness and joy. The entire country celebrates it because it is the birthday of our nation. As we look back to July 4, 1776 and see the immortal 55 signing the Declaration of Independence we hear them saying, "We must hang together or we will hang separately." They realized what it meant to them to sign that document and yet not one hesitated nor halted. They were brave men who loved liberty and were willing to sacrifice their lives for it.

Such men always win. The liberties of the masses have all come through sacrifice and usually a bloody sacrifice. The "Humble Nazareen" shed his blood that humanity might be free. He said, "I am come that ye might have life and that ye might have it more abundantly." The thirty years war of Europe was for religious freedom, and resulted in driving out of Europe to the bleak shores of eastern America, the most courageous, pure hearted and noble spirited citizens, and the founding of a new nation destined to be the greatest and grandest in the world.

It is not surprising that the children of the hardy, liberty loving pioneer settlers of America should throw off the tyrannical yoke of King George of England in 1776 and declare themselves free and independent. We would have been greatly surprised if they had not arisen and made themselves free. But it took a long eight years war with the mother country to make the Declaration an actual fact. It took blood shed. One surprising fact in the history of our country, is that in this home of the brave and land of the free, a race of people were enslaved for many decades and sold on the block like cattle and horses.

But the time came when the penalty must be paid and it was paid in blood. The immortal Lincoln said, in his last inaugural that, "He deplored the fact that the awful war of the rebellion still continued, but that it seemed necessary as a matter of justice, that as much blood must be shed from the veins of the white man in four short years, as he had drawn from the back of the black man by the lash during the past century." It took blood to wipe out the wrongs committed during this century.

We ought to appreciate our blood bought liberties. The thought should be uppermost in our minds on Independence day, how can we perpetuate the free institutions handed down to us by those who died for them. Our liberties are sacred because they are all blood bought. How shall we perpetuate the American Republic and make it the freest country in the world, where every person enjoys full liberty? This is a practical question for full discussion on Liberty Day.

Upon what does the perpetuity of our great Republic depend? You take a brick house well built, and on what does its strength and durability depend? You may say much depends on the architecture, the plan of the building, and much upon the workmanship wrought upon it, but all will agree with me, that the strength and durability of the great building depends primarily and chiefly upon the character of the individual brick which compos the building.

If the brick are rotton or crumbly, I care not how scientific the plan ofarchitecture or how skilled or honest the workmanship the building cannot be stable or strong. The same is true of the American Republic. You may tell me that the perpetuity of our liberties depends much upon the form of government and much upon our constitution, but you will all agree with me when I say that the durability and perpetuity of our free institutions depends chiefly upon the character ofthe individual citizens who compose it. If the majority of our citizens are ignorant and debauched and degraded the republic cannot be stable nor strong; it can not last. As our citizens are pure in their lives and intelligent and fraternal so the government in city, state and nation will be pure, efficient and for the benefit of every individual citizen. The idea follows logically then, that any institution which debauches and degrades the citizen is an enemy of the flag and not entitled to protection from old glory but ought to and will be destroyed by the American people.

Every organization like the church, school or Farmers' Equity Union, which promotes the character of the individual citizen is a friend of the American Republic and worthy of loyal support by all who love the flag. The by-laws of the Equity Union state that "Our object is to promote the morality, intelligence and fraternalism of our members and to make them golden rule co-operators." We are developing in our Unions men of the highest character of whom no organization need be ashamed.

The Farmers' Equity Union stands for economic freedom for all the people. We want to see the idea of "equal rights to all" carried out in the business world, but we believe that economic freedom like all other freedoms is only possible to an intelligent moral, fraternal people. As we exalt the character of the Equity Union members, we build up and make strong the grandest industrial Union in the United States of America.

MARKET REPORTS AND EQUITY PRICES.

The Farmers' Equity Union is a big question mark, continually asking good reasons for the continuation of present conditions on our farms, in the central markets, and in our government, or persistently pleading for a change for the better. We are neither standstillers nor standpatters, but believe in advanced, progressive methods of farming and marketing.

We think our backwoods system of marketing is responsible for more of the evils of farm life than all others combined. The farmer picks up the daily paper and scans the market reports for prices in our central markets. He finds columns devoted to the speculative market and about one inch to cash grain. The speculative feature should be eliminated entirely as a curse to humanity. The bulls run prices high on consumers when farmers have very little to sell, and then all turn bears when farmers are marketing a bumper crop.

We do not believe in price boosters on consumers, and price hammerers on producers. Why should farmers be compelled to scan the daily papers every day for prices on farm crops? Do they take up the daily paper for prices when they wish to purchase a wagon, binder, mower or gasoline engine? Why do they not scan the market reports for prices of farm machinery or anything else they buy? Because the farmers' products, alone, are bought and sold in a speculative market. His prices are made the football of gamblers and speculators. His markets are manipulated by members of the Boards of Trade, who often manage to have all our leading markets the lowest when the bulk of each farm crop goes on the market.

The mob of farmers also conduce to these bad conditions by their present dumping system so that poor farmers who cannot hold their grain, cotton or stock always find the lowest prices when they market. How different when he buys! He need not look for daily quotations or fluctuations in our market reports on farm machinery, boots and shoes, clothing or anything else he buys. Speculators have no power to manipulate or fluctuate their prices. They are fixed arbitrarily by a trust; a powerful combination which controls their prices having annihilated most of their competitors.

One million farmers must unite in the Farmers' Equity Union and have uniform prices in all our central markets each month. If it is right for manufacturers, wholesalers and retailers to have uniform prices, why not farmers? Grain, cotton, stock, and farm crops ought to be just as high when the poor man markets as any other time. When thoroughly organized, farmers will not scan the daily papers for market reports. They will look in the Equity column of their weekly agricultural paper and know the price for that month. A National Union is the only hope of the American farmer.

Golden rule co-operation as taught by the Farmers' Equity Union will unite the farmers and keep them united. Our plan helps every poor farmer who is honest and industrious. We give him the market when he threshes his grain. We will make the receipts light enough at threshing time to give the poor man one dollar a bushel for his wheat and a fair price for every other product.

In our Equity Exchanges we allow every farmer to come in on easy terms and have the benefit of co-operation. Our principle is equal opportunity, equal rights and equal votes or control. No man can own over four shares, and no man can draw out anything for patronage or interest till he has four shares, the limit. Then all are finally equal in control. The man is the unit all through. We measure men by their honesty and industry and not by their dollars. We respect the manhood of every individual and teach him to respect himself as a man. Money is not to rule one minute in an Equity Exchange, but men. Plutocracy, aristocracy, moneyocracy, and old money bags are all kicked in the ditch, and the plain, common people (God's kings and queens) are to rule in business, politics, society, everywhere. They are being educated by the Equity Union. We want an Equity meeting the first Saturday of every month regularly all over the country. Country teachers and editors must assist. Our Exchanges must run on safe margins, have good business management and never declare over 5 per cent dividends on the stock subscribed. Then some money can be paid back to each stockholder each year for his patronage which will show him the difference between the old and the new system. Most farmers must be shown in dollars and cents. They are like children. They want to count the money. Always buy on a safe margin and never declare over 5 per cent dividends. This is fundamental. Never vary from these principles. We must unite the farmers. Our goal is a National Union of farmers. Here lies our most wonderful possibilities. A National Union of farmers would have made the 1906 grain crop bring the farmers three hundred million dollars more than it did. Instead of losing millions of dollars on the hog crops in 1907 and 1908 they would have been sold at a fair profit. A National Union will guarantee the farmers a profitable price on every product in the future. Our co-operative system if made national in scope, will prevent the robber profits now made off of farmers annually, to the extent of hundreds of millions of dollars. But the stone-masonry work must be done for this grand structure first. This is imperative, fundamental. Honest, pure-hearted, clear-minded lecturers must be put in the field and kept hammering away at the walls of ignorance, suspicion, prejudice and selfishness which now separate and

Sledge-hammer blows must be dealt against their carelessness, indifference and inaction. The rising generation will be educated to be co-operators. This is our greatest hope. This buoys and encourages every co-operator. Our labor is not in vain. Where we sow the good seed others will reap the harvest, for the harvest is sure to come. Be not weary in well-doing, for in due time we shall reap if if we faint not.

FIFTY-SEVEN HAY DEALERS IN CONTROL

The Hay Dealers' Association is an organization of middlemen. They are organized for protection. They want their business to be sure and profitable. They want every individual farmer to give them enough out of what he produces to make a good living for every one of them and their families, and so they organize. They get when they take a notion they are not getting enough from the farmers, they meet together March 7th, 1914 and decide to increase the close together in the "Kansas City Hay Dealers' Association," and weaken our farming fraternity. Farmers must be aroused to action. Commissions 50 per cent. Kansas City, being the largest hay market in the world, this raise of the price for selling hay will mean a heavy loss to the farmers who produce the hay. Now what is Mr. Individual Farmer going to do about it? He must sell in this organized market or do without a market for hay. He has no protection whatever without a Union.

FIFTY-SEVEN ORGANIZED DEALERS HAVE FULL CONTROL OF THE THOUSANDS OF UNORGANIZED FARMERS ALL AROUND THE KANSAS CITY HAY MARKET. The biggest hay farmer in Kansas has no protection. He is an individual and the dealers are an organization and have the equipment, capital and control of the market where he must sell.

His only hope is to join the Equity Union and help us build it strong all around Kansas City and then establish our own Central Hay Market in Kansas City. If the farmers all around Kansas City will join the Equity Union and build it up strongly, we will have our Central Hay Market and instead of raising prices for handling, lower them.

PURPOSES, PLANS AND PRINCIPLES

The name "Equity Union" indicates our purpose and character and principles. Equity is the basis of all righteousness, and Americanism is the creed of every patriotic citizen of our republic, hence our union is sure to appeal to all thinkers and lovers of liberty. Webster says equity is giving or desiring to give each man his due. Equity granted and demanded is a brief definition of our position as a union in all relations of life.

The man who really and honestly believes in equity, would as soon be oppressed at to oppress others. Our movement only hurts those who live by prey and their allies who for profit practice inequity. We make no distinctions in the rights of men on account of politics, religion or honorable calling. This would be un-American. We fear no exposure, and dread no test. Our purpose being just, our demands fair, and our methods honorable, we are more than willing that all shall know us for what we are in fact—The American Square Deal Association.

Man being a land animal, his life depends on the products of the soil. Without these he cannot live, much less prosper. This being true, the most important business in the world is farming. The foundation of all industrial and commercial structures is agriculture. The farmer is therefore the most important factor in the prosperity of nations, and ought to be the chief concern of government. When crops fail, all suffer. Those who attack his interests, wound themselves. Every human being engaged in legitimate business ought to be the farmers' friend.

But the farmer is under no moral obligation to feed and clothe the world without a just reward for his labor. He has the natural right to name the lowest price at which farm products shall sell. This right is claimed by every man who makes a machine of any kind. If farmers have something to say about the price of their products the legitimate profits on their business will be retained by them as they should be. If gamblers, speculators or trusts fix prices on farm products, the profits on farming go to them, and the whole community in which farming is done is impoverished.

Self-Evident Truths.

Good prices and good times are inseparable. Low prices and hard times go hand in hand. We hold this truth to be self-evident that the farmer who farms his farm has a better right to fix the minimum price on the fruits of his toil than has the speculator who farms the farmer. The Farmers' Equity Union is not a trust. The chief difference between a farmer and a trust magnate lies in the fact that the trust magnate works everybody for himself, and the farmer works himself for everybody. Trusts are organized to restrain trade. Farmers are organized to unfetter trade. Trusts are combinations of the few for the few. Our union breaks the power of combinations between producers and consumers, which now rob both classes. Increase the purchasing power of farmers by sure, equitable prices and they will buy more of the products of mines, mills and factories. This will stimulate all industries and spread prosperity to all classes.

To know the supply and demand and fix a just price that will insure the farmer a just reward for his year's labor is part of the mission of our union. No movement ever involved greater possibilities for good, and few organizations of men have a higher mission. Farmers have always had prices fixed for them both ways. From the beginning of time they have stood in the markets of the world, asking two questions: "What will you give? What will you take?" Once each year the farmers have in their possession everything which makes life on our planet possible, and then let gamblers, speculators and middlemen take it away from them at their own price. After it passes into their possession, prices go up and they absorb the profits. This is not good for the farmers and consumers. They make farmers take less than their due, and compel consumers to pay more than is just.

Age of Organization

This is an age of organization, and unless farmers unite, their rights cannot be maintained. The whole scheme of marketing, distribution and transportation, has been rigged for the purpose of taking from the farmer the millions his toil has produced. Unless producers have the intelligence and courage to organize for mutual protection they will continue to be robbed.

Farmers do not work on commission nor for salaries. Their income depends on prices. Organization and co-operation is therefore absolutely essential. Farmers plow and sow by faith, live by hope and market by accident. In the absence of intelligent organization and co-operation, farming is a game of chance. Having no power (except through organization) to prevent low, unjust prices, the farmer never knows what the result of his year's work will be. In case he raises a big crop and "dumps" it on the market, he will be in luck to realize the cost of production and enough to "winter him over" for the next season. Should adverse conditions result in a short crop, the surplus he "dumped" the preceeding year will be used to hammer down prices for his new crop until he parts with it, after which, the speculators will boost the price and rob the consumer.

Before knowing what the yield will be, many farmers are compelled to mortgage and contract their crops in order to live. Thousands who feed and clothe the world work all their lives for their board and clothes. Millions of farmers live one year behind the procession. The ledgers of merchants and mortgages prove this. In some of our most prosperous dairy counties farmers are borrowing money at the banks to pull through the winter while the millionaire condensing company is building a tenth new plant in another state out of the profits from this mob of unorganized farmers.

Shall these conditions continue? Not unless they are continued by the farmers themselves. The solution lies in a National Union of farmers founded upon golden rule co-operation. Price making is a matter of feeding the market. As the Equity Union proposition is to fix bottom and not top prices, success is absolutely certain, provided farmers join hands in honest, earnest effort to market their crops.

Dump and Demand

Of course your mind naturally reverts to the natural law of supply and demand, which is supposed to regulate prices. People who rely on this law should explain how packers often manage, under its operation, to pay farmers four cents per pound for fat hogs, and charge consumers twenty cents per pound for bacon. How could a short crop like the 1911 wheat crop all sell under the dollar mark while flour and feed soar sky high?

In studying this law farmers have overlooked the important fact that it is what **they do with the supply** that determines prices. We must feed the market! When farmers organize and learn to do this, prices on all farm produce will be as steady and equitable as on farm machinery. Then only those who sow shall reap. Under the present system, those who neither sow, nor reap, gather the richest harvests. No wonder grim want lurks in the shadow of gilded palaces within whose walls rich criminals revel in luxury.

Victory Through Organization

When farmers organize and agree that wheat shall not be sold for less than one dollar per bushel, the entire speculative world in self-defense will be compelled to support their contention. Neither bulls nor bears can force the price below one dollar when the wheat growers unite.

WE WILL BUILD OUR OWN FACTORY

The factories are afraid of a farmers' union. **The owners are capitalists.** They don't believe in co-operation for "the people." They want co-operation for the capitalists—**not for the people.** The whole business world is capitalistic in principle and practice.

We will never get the price and quality right for our members on autos and wagons and farm machinery until we organize an Equity factory of our own. But we must organize a **market for our out put.** A market that cannot be bought by the machine trusts.

We can get plenty of capital and patrons through the Equity Union if all of our members will go to work. A good living price will give us the very best skilled labor. We will want the very best material that can be bought. We want our factory to run at full capacity and we want to know that our machines are sold. Then our Exchanges will have the warehouses and distribute direct from our factory to our members. We can make the best eight foot self-binder and deliver it away out at Rhame, N. D., or at Tyrone, Okla., for $100. There is no doubt of this but it will take a strong National Union of farmers **educated** to follow Equity Union principles. We must have one hundred thousand **educated co-operators** in our Union, who will own only $20 each of the capital and never declare over 3 per cent stock dividends. Then these members must all buy from their own factory. There are many localities near Kansas City where we could deliver our binders for $75 and guarantee quality and repairs. If we can run a factory successfully and mills too, it will not take long to unite one million farmers and reduce the cost of living more than 25 per cent.

THE PROFIT-TAKERS ARE IN THE SADDLE

The Profit-Takers are having their inning. They run the business of our country for the people? No! For themselves! They use every pretext for raising prices on a long suffering public. The European war comes along and up go prices, sky high. In some cases the raise is legitimate, caused by natural law of supply and demand, but in many cases, they use the war cry as a pretext for gouging the people. There are many honest business men and we are not condemning them by the wholesale, but there are so many greedy, selfish middle men and they control the business situation and put prices too high whenever there is a pretext, so that they dare to do it.

Our retail and wholesale business, as well as our manufacturing business, are run on an entirely wrong system and always will be as long as we allow profit-takers and grafters and extortioners to run our business for us. The business idea of today is to buy as cheap as possible and sell as high as possible.

This is not the idea of every business man. Many of them are trying to give good service for the money they take from the people. But the prevailing idea among many middle men is, "make money boys, make money honestly if you can, but **make money boys!**"

The people must be aroused to the fact that "our business" is not in safe hands. As we are beginnig to wake up politically and wrest our government from the grafters, so must we organize the wage-earners and farmers into industrial unions which will take complete control of the business of this rich agricultural country. We must educate the people to be golden rule co-operators at heart. Then they will unite and change the business system of our country. **The men who run the business of our country for profit will not run it for the people but for the few.**

Graft and greed are bound to control our business system as long as we allow profit-takers to control it. The Equity Union is on the right track. The little Giant is marching on and conquering fort after fort of the enemy. There is wailing and gnashing of teeth among the profit-takers where we are organized but rejoicing among the common people who work for their living. At 100 good markets the selfish, greedy profit-taker sees his doom written in letters of fire.

At 100 good towns we are demonstrating that golden rule co-operation is not only right but practical in business. In 1913 we saved one million dollars from the unholy coffers of the profit-takers and spread it out among the rightful owners, who were members of the Equity Union. This was accomplished by the Equity Union, notwithstanding the fact that nearly one-fourth of our members who received benefits from the Union refused or neglected to pay dues in 1913 and again in 1914.

We want every Equity Exchange to pay the national dues of each stockholder for 1915 as soon as possible and charge it to his account. The by-laws provide for this. Nothing wins like success. The Equity Union is a success. We are demontsrating that our plan of co-operation unites the farmers and makes them golden rule co-operators. Our business system will overthrow the robber-profit-system and bring the economic millenium. If "the people" were in control of their own business in the country and central markets through co-operation, would they put prices up sky-high on themselves?

Let every man who can, reason and then think hard on these things. The Profit-Takers are in the saddle but they will be unhorsed and overthrown in time by Equity Union golden rule co-operation. Brother farmer, go after one more member. Get your neighbors to read this paper. If you will get them to put 50 cents into this paper and read it one year, they will join the Union. Try them!

FORTY EQUITY UNION RALLIES A SUCCESS IN DAKOTAS

The glad welcome accorded to me by the people in forty big Equity Union rallies in North and South Dakota during June and July was the biggest reward I have received as president of the Equity Union for three and one-half years. The pay received in money has not been large but ample. The hearty and full approval by the people of the service rendered to the farmers, through the Equity Union, is the richest and most satisfactory reward I could receive.

I went into the Dakotas in 1911 with strong faith in the Equity Union principles and business plan and determined to get the farmers to unite and carry out golden rule co-operation in business. I worked hard for four months on the Milwaukee railroad and received curses and bad names of all kinds from some farmers at every town where I worked. But at every place I organized a good many intelligent progressive farmers, who were willing to work with me to educate their community up to the high standard of golden rule co-operation. When I went back in the summer of 1912 and found a good crop in sight I was able, through the help of many good reading, thinking members, to organize a few Equity Exchanges. All of these made demonstrations showing that the Equity Union plan of marketing is the right way for the farmer.

The organization grew steadily in 1913 in these states and the forty rallies this year where thousands of the people have gathered as Equity Union members, show conclusively that the people in western North and South Dakota are thoroughly aroused against the profit-taking system, which robs poor hard working people for the benefit of the rich. The many hearty hand shakes and rousing cheers of large audiences in these forty big meetings have made me rejoice more than a ten thousand dollar salary would do. The triumph of a great principle through the Equity Union is sufficient to cause any man to rejoice who loves his fellowmen and hates graft and robbery of the common people.

We take this opportunity through our grand little paper to thank the thousands of men, women and children of North and South Dakota for the glad welcome you gave me in the forty big rallies in June and July. These large enthusiastic meetings mean much to our cause wherever we are started in the United States. They have strengthened the faith and courage of all of our members and made hundreds of converts to our cause. Fifteen new Equity Exchanges will handle the fine 1914 crop and save more than seventy thousand dollars to the members, and yet hundreds of our members are holding back the 1914 dues. If every member had paid 1912-13 and 1914 dues we would have been twice as strong as we are. Many of our good workers have left their own business and paid their own expenses to work for Equity Union until they were not able to go further with the work. Some of our Exchanges have allowed men to take stock and reap benefits without being members, but the forty big rallies in the Dakotas have cleared up the atmosphere. All understand clearly now that we can never build a Union that way.

Members' dues must be paid from the Exchanges. Only members of the Equity Union must be allowed to take stock in an Equity Exchange. Outsiders must not be allowed benefits till they become members. Every farmer can become a member and help us work out golden rule co-operation for himself and others.

The conference meeting in Mott in June did much to solidfy and establish permanently the Equity Union on the western slope.

The organization of the Aberdeen Creamery and Mercantile Company will make Aberdeen an Equity Union center. Here we will get our central market demonstration for the Northwest. Unions will be organized in every direction on the ten railroads running in from ten directions. There will be a great advantage in having a good center and organizing solidly all around it. The questions of transportation and distribution enter largely into our problems of co-operative buying and selling. We want to again thank the thousands of Equity Union men, women and children for the welcome extended to us in the forty big rallies in the Dakotas in June and July 1914.

RHAME, NORTH DAKOTA

Saturday Sept. 5th, 1914, Rhame Equity Union held its first monthly meeting since the organization of an exchange and purchase of an elevator. Threshing was just reaching the rush stage over our territory and to say that we are pleased with the size of our business with our manager and with the certainty of a big saving to our members but mildly expresses our satisfaction.

Our board of directors and manager have laid down the following rules of business which will be strictly observed.

First—the absolute correct weighing and grading of every bushel of grain delivered at our elevator.

Second—exactly the same treatment of every man whether he sells us 10 or 1000 bushels.

Third—the buying of all grain at as small a margin of profit as is consistant with a safe and solid business management, these profits, of course, to be prorated back to the stockholders at the end of the year, instead of being handed over to Minneapolis and Duluth grain gamblers, together with a lot of other profits collected by them in various and devious ways from the farmers. An examination of our business to date shows that our organization has raised the general price of grain here from two to five cents per bushel exclusive of any patronage dividend in the future.

It shows that we are paying within one cent per bushel of the price paid at Aberdeen, S. D., and I understand that grain freight rates from Rhame to Minneapolis are five cents more than from Aberdeen to Minneapolis, which gives our organization credit for 4 cents to all grain raisers in addition to the patronage dividend to stockholders.

We are constantly selling more stock as farmers come to understand and realize the special benefits to stockholders and we confidently expect before the end of the year to have all of the progressive and thinking farmers on our stock list and boosting with us for the success of the best and most practical plan that I have ever seen for correcting and preventing the shameful abuses now being practiced in the buying and handling of the farmers' grain. We learn that in a few instances an old line elevator has, in order to bait a farmer, dishonestly boosted the grade by giving his grain a higher grade than it would stand, thus adding 2 to 4 cents to the price. But I want to remind all farmers who are so baited, that the fact has leaked out that neither the old line companies or any other company that I can now think of, is in business either for their health or for sweet charity's sake, and whenever they make some farmer a present of a few cents or dollars for a bait they immediately steal enough from some other farmer through dishonest grading or otherwise, to make up the loss. It is a notorious fact that this is their plan of business and if we take the bait and continue to support them and pour our hard earned money into their treasuries today, we may be the man they beat in order to bait another farmer. On the other hand, in our organization there is absolutely no reason or temptation to dishonest weighing, grading or other questionable practice. In order for a farmers' elevator to steal for themselves they must steal from themselves. Co-operation is the slogan of every booster and farm paper in the United States and Europe. We have established in Rhame, the foundation for a co-operative business along the best and safest lines that I have ever seen. Here is an opportunity to practice what we preach and what the best brains all over the world is preaching, to wit: Co-operation among farmers. Let us get together, all join the Union, each member buy at least one share of stock. Boost for the Union and stick by it. That is the only way we make a success of any business by sticking to it. And why not? If we do this, we all know to a certainty that at the end of the year, when the balance is struck and we have done our own business and kept and divided our own profits, that each and every man has got every cent there was in his crop instead of handing over a big slice of it to some foreign grain company.

At the time of buying the elevator business, I confess that some of we directors felt a little shaky on account of the extremely discouraging outlook for crops. But prospects have changed, and with the present high prices and at least double the yield expected, I believe our territory will be in a good shape this season financially, as the average year, if not better and a good profit on the grain business is assured, so again I say, let's get together and stick together. In no other way can we make so good a profit on so small an investment.   F. J. WEIR, Pres.

HAYSEED AND EQUITY

Hayseed: Hello! Equity.

Equity: Good morning Hay. I came over to see if you would not join our Union and haul your wheat to our Exchange.

Hayseed: No, not while Old Line pays 2 cents a bushel more than your Equity Exchange.

Equity: Do you know what he did two years ago, when we had no Exchange?

Hayseed: I know he was away down that year.

Equity: I know I hauled right by his elevator to the next town, where there was a farmers' elevator and made 5 cents a bushel on my wheat and I know he will never get another bushel of my grain for any price.

Hayseed: Well, I go where I can get the most money for my crop.

Equity: If the Equity Union was not here, would he pay 2 cents a bushel above the market?

Hayseed: No he would be lower than he is now, sure.

Equity: If you and others help to kill the Equity Exchange for 2 cents a bushel, how much will you gain by it? Old Line will have control of our market again and make that 2 cents back and another 2 cents with it. You ought to know better. Don't support such a system. The sooner Old Line is out of business the better for every farmer. Last year I stuck to Equity Exchange when the enemy was paying 3 cents above the market and my proration gave me 4 cents back. I did'nt lose 3 cents but gained one cent a bushel. I want a market where I can send my boy with a load of wheat and I will get a square deal on weight, grade and price any and every year.

Hayseed: Neighbor, I believe you are right! I am going to join the Equity Union and haul every bushel of my grain to the Exchange. I don't care what he pays. He can't buy me nor my grain.

RESOLUTIONS

The following resolutions were unanimously adopted at Mott, North Dakota in the Equity Union Convention June 22 and 23, 1914.

Resolved that it is the unanimous sentiment of this convention that only members of the Farmers' Equity Union have a right to take stock in an Equity Exchange and that we demand that every Equity Exchange in the Dakotas purge itself of enemies of the Equity Union. They must either be loyal members or get out.

Resolved, second, that every Equity Exchange must pay the national dues of each stockholder and charge the same to his account on the books, in accordance with the By-laws.

Resolved, third, that no Equity Exchange has a right to declare over 5 per cent stock dividends, in violation of the national By-laws adopted and ratified by our three national Conventions at St. Louis, Chicago and Kansas City.

Resolved, fourth, that the manager and directors of each Exchange are to be the judges of what is a safe margin in buying and selling, but we as a convention are opposed to an unsafe margin at any Exchange.

Fifth. Every member is expected to loyally support his Exchange with his patronage whenever the Exchange is able to handle the business.

Sixth. Every Exchange is expected and kindly requested to equip itself, as rapidly as possible, to handle the growing business which is being organized by the Equity Union at each market until every bushel of grain will pass through an Equity Exchange.

Seventh. It is the unanimous sentiment of this convention representing west from Bismarck on the N. P. that the Farmers' Equity Union of Greenville, Ill., has the most practical plan of co-operation ever adopted by farmers and that the steady growth of this young giant in twelve states and the thousands of dollars prorated to its members the past two years are evidence of this fact.

EQUITY UNION CONVENTION IN MOTT, N. D.

The first convention of the officers and managers of the Equity Union of southwestern North Dakota and Northwestern South Dakota, met at Mott on Monday and Tuesday of last week, the event proving a gala day for Mott and Hettinger county. Prominent speakers were here from both states as well as National President C. O. Drayton of Greenville, Illinois.

The interest manifested showed conclusively that the farmers and their representatives are alive to the situation, and the wide-spread movement for establishing a sound market for the grain of our country, is fast approaching.

The meeting on Tuesday evening was called to order by Pres. Drayton. J. P. Larson of Mott was elected as secretary of the convention by a unanimous vote. The convention immediately got down to business. It was moved and carried that a committee of three be appointed to draft resolutions to be brought before the convention on Wednesday, relating to the "Co-operate laws" and also the "reciprocity demurrage."

The following named committee was appointed; E. N. Bosworth, A. K. Moehn and John Stephens. It was moved and carried that enrollment be made of the delegates from the different unions. A committee of three was appointed by Pres. Drayton on order of business for Tuesday as follows: Messers. Hoffman, Bratsberg and J. P. Larson. After appointing of various committees timely discussions were taken up in which many of the delegates participated, after which the meeting adjourned until Tuesday morning.

On Tuesday morning, Pres. Drayton called the convention to order shortly before nine o'clock, the Family Theatre building being packed to the limit. First business transacted was reading of rules and order of business, by the secretary, which were adopted. It was moved and carried that an expert salesman be employed at a future date, just as soon as the exchanges are ready for his services. It was further moved as the sentiment of the convention that an Equity Union Creamery and cold storage be built at Aberdeen, S. D. This motion carried unanimously. Moved that a committee of three, to take the matter up with the National Union to confer and work with that body for a large central creamery. This motion was lost.

It was moved that each local union be asked to appoint one delegate to be sent to Aberdeen, to convene in session at a time set by the National Union. These delegates to be appointed by each local at the regular meeting on the first Saturday of July. It was moved that the reports of the committee on laws be adopted. This motion carried. Moved that this convention appoint a committee of three to arrange for a uniform set of books for the different exchanges. Messers. F. G. Orr, Moehn, and Stinger were appointed. It was moved and carried that this convention endorse the Kansas Co-operative law and that a committee of three be appointed from each local to see their representatives in district and request them to work in behalf of the passage of a similar law for our state.

Moved that this convention tenders its thanks to the citizens of Mott and also the National Union for the many courtesies shown. It was moved and carried that adjournment be now taken. President Drayton addressed the gathering several times during the day's proceedings, while many of the delegates gave informal talks.

At the close of the meeting, citizens of Mott with autos were in readiness at the convention hall with autos. The delegates were given a trip through the country and given an opportunity of viewing our community in all directions. The Mott concert band rendered a pleasing concert during the afternoon, their efforts being highly appreciated by the visitors.

J. P. LARSON, Secretary.

A REAL CO-OPERATIVE LAW
CO-OPERATIVE ASSOCIATIONS.

Any number of persons, not less than twenty, who are citizens of the state of................, may associate themselves together as a co-operative corporation for the purpose of conducting any agricultural, dairy, mercantile business on the co-operative plan. The title of such corporation shall begin with "The" and end with "Association," "Company," "Corporation," "Exchange," "Society," or "Union." For the purpose of this act the "Co-operative Plan" shall be construed to mean a business concern that distributes the net profits of its business by: First, the payment of a fixed dividend upon its stock, not exceeding 5 per cent; second, the remainder of its profits are prorated to its several stockholders upon their purchases from or sales to said concern or both such purchases and sales. They shall sign and acknowledge written articles of incorporation which shall contain: The name of the corporation; the names and residence of the persons forming the same; the purpose of the organization; the principal place of business the amount of capital stock; the number of shares and the par value of each share; the number of directors and the names of those selected for the first term; the time for which the corporation is to continue, not to exceed fifty years. The original articles of incorporation or a certified copy of the same shall be filed with the secretary of state who shall return to the corporation a certified copy of the same, with the date of filing and attested with the seal of his office, upon the approval of the Charter Board. For filing the articles of incorporation and amendments thereto under this act the same fees shall be paid to the secretary of state as is now required under the general corporation law. No corporation organized under the provisions of this act shall commence business until at least twenty per cent of its capital stock has been paid for in actual cash, and a sworn statement to that effect has been filed with the secretary of state, and his receipt for the same shall be construed as a permit to do business. Every such association shall be managed by a board of not less than five directors. The directors shall be elected by and from the stockholders of the association at such times and for such term of office as the by-laws may prescribe, and shall hold office for time for which elected and until their successors are elected and shall enter upon the discharge of their duties; but a mjaroity of the stockholders shall have power at any regular or special stockholders' meeting legally called, to remove any director or official for cause, and fill the vacancy, and thereupon the director of said association shall cease to act. The officers of every such association shall be: a president, one or more vice-presidents, a secretary, and a treasurer, who shall be elected annually by the directors, and each of said officers must be a director of the association. The office of secretary and treasurer may be combined; the person filling the office shall be secretary-treasurer. No person shall be allowed to own or have any interest in more than ten per cent of the capital stock of such corporation. Each member shall be entitled to one and only one vote for each director to be elected. Each corporation shall formulate by-laws prescribing the duties of the directors and officials; the manner of distributing the profits of its business; the manner of becoming a member; and such other rules and instructions to its officials and members as will tend to make the corporation an effective business organization. Each corporation organized under the provisions of this act shall make an annual report to the secretary of state the same as required of other co-operations; provided, such co-operative corporation shall be required to report the names of its stockholders and the amount of stock owned by each for such years only as may be required by the secretary of state. All co-operative corporations, companies, or associations heretofore organized and doing business under prior statutes, or which have attempted to so organize and do business, shall have the benefit of all of the provisions of this act, and be found thereby on paying the fees provided for in this act and filing with the secretary of state a written declaration signed and sworn to by the president and secretary to the effect that said co-operative company or association has by a majority vote of its stockholders decided to accept the benefits of and be bound by the provisions of this act. No association organized under this act shall be required to do or perform anything not specifically required herein, in order to become a corporation, or to continue its business as such. No corporation, association or company now or hereafter organized or doing business for profit in this state shall be entitled to use the title "Co-operative" as part of its corporate or other business name or title, unless it has complied with the provisions of this act; and any corporation, association or company violating the provisions of this section may be enjoined from doing business under such name at the instance of any stockholder of any association legally organized hereunder.

A CO-OPERATORS' LAW

"The People" are awakening to the fact that golden rule co-operation is their salvation in the economic world. But the capitalists have been in full control of our government for a century and the laws are nearly all made in their favor.

In Ohio, Illinois, Iowa, Missouri, and Oklahoma, the law is made for the corporations, but not for "the people," who want to be golden rule co-operators.

The law in North Dakota is not what it ought to be. We ask our members in these states to read carefully the co-operator's law in this issue of our paper. This is the law we want enacted in the six states, North Dakota, Illinois, Ohio, Missouri, Colorado and Oklahoma, where we are organizing co-operative Exchanges.

The other states where we have Exchanges have fair co-operative laws. This law recognizes a co-operative Exchange and makes its provisions legal in the courts. It defines a co-operative plan and shows "the people" what is meant by true golden rule co-operatio. **It will be an educator.**

"No corporation, hereafter organized or doing business for profit in this state shall be entitled to use the title "Co-operative" as part of its corporate or other business name or title, unless it has complied with the provisions of this act." They must not sail under false colors! The business world is full of fakes and humbugs.

This law prohibits over 5 per cent stock dividends and gives "the people" the right to prorate the net earnings according to patronage. The man votes and not the dollar. Each stockholder has only one vote.

By-laws, giving the Exchange the right to exclude objectionable person can be **made.**

Now we want to pledge as many candidates for the legislature as possible to vote and work for this law, if they are elected. Send your paper to some candidate for your legislature and write to him and ask him if he will try hard to get this bill enacted into law if elected. Tell him this is the law the farmers want. Let every Equity Union member get busy and secure one pledge for this law. It is very important that we have co-operative laws in all of the states where we are organizing the Equity Union. Write to us when a pledge is secured and we will publish the name of the candidate. This move is entirely non-partison. We want fifteen or twenty members of your next legislature pledged to work for this bill and we will get it through.

STARVE THE POOR AND ENRICH THE RICH

Attorney General McReynolds and Secretary of Commerce Redfield have sent agents throughout the country for the purpose of investigating the sudden jump in the price of the necessities of life. The agents are directed to find out what articles are effected and the reason for the increased price and whether the cause is speculation. The speculators are so combined and equipped and capitalized that they can raise the price of the necessities of life and starve the poor people whenever they can find a pretext so that they dare do so. The indications are that the recent advances in price are due to speculative greed. The extortioner has always been despised by the people and will not be permitted to flourish in free America in the twentieth century.

The elimination of this class of fine haired gentlemen is the primary object of the Equity Union. Their insasiable greed for gold makes them inhuman to humanity. They must have big profits on the necessities of life even if millions of poor people suffer as a result. Their love for gold squeezes out all love for humanity. The **love of money is the root of all evil.** The extortioners and profit-takers have control of the business channel from producer to consumer. The question has been agitated in our country for fifty years as to how we could eliminate these gentlemen from the distributing end of all farm produce and make them go to work for their living as honest people do.

We believe the Equity Union plan of golden rule co-operation will work to this end more and more as the people are organized into this great and beneficient industrial Union. We have only made a good hopeful start in twelve states and yet the air is sulphurous in some places with the curses of some of this cloth of gentlemen who have been put entirely out of business. Let the good work go on. Our system of distribution from farm to consumer and from factory to farm is all wrong at present and must be changed by "the people," who suffer from it.

The good work of organization and education must go on. This is the only hope for the people. Every community must be made solid for Equity Union.. Its principle of golden rule co-operation must be instilled into the minds and hearts of men, women and children in our homes. Its practical business plan must be demonstrated. Then "the people" will unite and overthrow the "tables of the money changers," whose greed for gold respects nothing sacred.

Next to the church and school comes the Equity Union. These three great institutions are bringing the Millennium day. Overthrow the profit-system and Mr. Extortioner will go down with it and industrious, honest people will have a chance to live. This is our mission as a Union.

OUR COMRADE OF NAZARETH
By S. J. Duncan-Clark

The world is making a re-discovery of Jesus. It is finding the Man back of the creeds and theologies that have grown up around His name; it is feeling again the heartbeat of our Comrade of Nazareth under the cerement of dogma with which He has been bound about through centuries of religious scholasticism. So many and so intense have been the controversies waged around Him that in the conflict of opinions, He himself has often been obscured. Men have spent years fighting for some article of creed concerning Jesus that might have been spent as His comrades, fighting for a better world.

But Jesus is emerging from the clouds of theological dust. His figure is becoming clear to the eyes of men, and as the dream of social redemption takes to itself new beauty and definiteness men turn to Him for inspiration and leadership. We are hailing Jesus with a new enthusiasm today not because we think more of Heaven, but because we set a higher and truer value upon earth; not because we are more iterested in the mansions above but because we are becoming greatly concerned about the tenements in the next block; not that we may have a closer fellowship with angels but that we may have a better fellowship with one another.

If there is one note of music that rises stronger and sweeter above the tumult of life today than any other it is the clear, fine note of human service. To serve one's country, to defend the honor of its flag, to die with face set toward its enemies—these splendid purposes evoked the courage of men in days gone by, and still evoke them; but today a higher purpose is controlling many lives—the idea of service for one's fellows. Patriotism broadens beyond national boundaries. We realize that God has no chosen race, no favored people. In the glad comradeship of Jesus we claim the world as God's country, and every man, whatever his creed or color, as brother.

And Jesus, as the personal embodiment of this ideal, enters into every phase of life today. There is no need to await the call to arms, the unfurling of the colors, the donning of uniforms and the blare of martial music; there is no need to turn aside from the common tasks, the daily vocation, the duties of home and office and factory, that we may share His comradeship in the service of this ideal.

To tarry for the great occasion is to miss the immediate opportunity. The culminating tragedy of Jesus' life was reached by a pathway of ceaseless service for men in the little matters of every day. The cross derived its significance from the career that was crowned at Calvary. It would have had small meaning, indeed, were it not for the carpenter's shop in Nazareth, the ministry in Galilee, in Samaria and in Judea. Because Jesus touched so closely and so helpfully the lives of publicans and sinners in the streets of Capernaum, of Bethsaida and Jerusalem, He touched the life of the world when transfixed upon Golgotha. Back of that dark hour which made succeeding centuries luminous with love were the smiles of little children, the rested hearts of the wayworn, the strengthened hands of those who were fainting beneath life's burdens.

Nor is it in the slavish imitation of Jesus that the secret of comradeship lies. Rather it is in catching His spirit and getting His vision that you can become His comrade. Only as He helps you to find yourself can He help you to serve your fellows. The cause for which He lived and died needs more than anything else the investment of personality. It is this investment from which so many shrink. It is for the lack of this investment that religion falters and fails and loses its grip upon life. No measure of mechanical, card-indexed, scientific philanthropy can compensate for its absence, however rich the financial endowment.

Personality is the one asset that every man possesses, to do with as he pleases—and it is the asset of greatest value. Catching the spirit and vision of Jesus, quickened by the sense of human brotherhood, inspired by the dream of justice, you can put personality into every relationship of life. You can readjust the day's work to this new motive, you can make it the basis of companionship. You can live to its music in the crowded street, the public mart, the quiet circle of friends or the turbulent arena of politics and affairs. And the investment of personality means the increment of personality. It is thus that we grow and broaden and become enriched; it is thus, in the comradeship of Jesus, that we find ourselves.

LOCAL CLUB OR NATIONAL UNION

One of the many very good farm papers covering this Northwest field is very earnest in its advocacy of the farm club movement but is inclined to doubt the wisdom of organizing national societies of farmers, having a national board of control and requiring the payment of entrance fees and annual dues.

The writer is a member of one of the first farmers' clubs organized in Hettinger county and is a steadfast friend of the better farming propaganda. The field men of the Better Farming Association who have worked in this country know that I have been a friend to them and their work. But this is not relevant except to show that what I may say about farms clubs is not said with a purpose of knocking this institution. I am also a member of the Equity Union and it is in defense of this organization that this letter is written. Not that the Union has been assailed, but that it seems necessary to dispel the idea prevalent in some quarters that the farmers' club is the panacea for all our woes and that such organizations as the Equity Union are unnecessary and impracticable.

The purpose of the farmers' club and of the Equity Union are different in the main idea and yet quite similar in minor points. One striking difference is in their inception. Since the clubs of North Dakota were organized and are fostered by the B. F. A. movement. Then the club may be said to be an uplift agency or rather, the suggestion comes to us from without that we should begin to do a little uplifting ourselves. They agree to furnish a man to show us how to take hold and encourage us with an occasional "heave, O heave" Personally, I haven't the heart to spurn their well-meant efforts to help us. Their efforts are not only well meant but well directed and I think we should make the fullest possible use of them. Meanwhile, in the Equity Union, we shall be learning another lesson, so that when we get to be "better farmers" and able to show a larger gross income our benefactors shall not be able to garner more than their equitable portion of it.

The question whether we shall have farmers' clubs or a national organization of farmers may best be answered as the little boy answered his mother when she asked whether he would have a piece of the gooseberry pie or a piece of the apple pie—"both, if you please." We should have both the farmers' club and the Equity Union with its National Head and subordinate unions. The local farm club may unite a whole community, more or less remote from town, in a social, educational and financial uplift campaign. The club teaches community co-operation, tolerance of others' views on matters of politics and religion; it trains the farmer to think about his business and to express his thoughts to his neighbors; it condenses and crystallizes into usable forms the sum of neighborhood knowledge on all farm subjects; it unifies the community and makes it a better place to live; it furnishes a wholesome outlet for youthful energy which might, otherwise, be expended in less innocent and less helpful ways. It encourages musical talent; it discovers the orator; it develops leadership. In view of the fact that the farmers' club may do all these things and a great many more it is hard to see how we could give it up. But after all these things have been done, or have been begun there is still a large field of usefulness open to such organizations as the Equity Union. As previously stated, the farm club idea, as we have it in North Dakota, came to us from without, but the Equity Union had its origin in our own household; it is the child of our own brain.

The originator and organizer of the Equity Union is a farmer. This same farmer is our national president. The Union has a more daring and ambitious purpose than the farmers' club. It presumes to teach the farmer how to work with his neighbor, rather than the error of the old idea that the farmer is independent and undertakes to show that every tiller of the soil is dependent upon every other man similarly engaged; that "no man liveth to himself alone;" that, standing alone, the farmer is helpless and at the mercy of an army of parasites who, from long feasting, have become too strong for him to shake off by his individual effort. It dares to assert that the farmer is entitled to all the fruits of his own labor; that the piratical hands which hitherto have been permitted unquestioned, to take toll from his produce as it passed from the farm to the consumer, shall be withdrawn and be employed in productive labor.

The co-operative handling of the everyday necessities of life is the red thread and which runs through every Equity Union enterprise. To save money for the farmer was, I think the original idea of our organizers. The Union theory is that the best place to begin better farming is not at the beginning but at the latter end of the process, namely, in the marketing of the crop. Better business methods in selling his produce will give the farmer a large net income. But what could a farmer do with that extra income? He can buy new and better machinery so that next year he can do better farming in the sense of better tillage and higher production. He might buy another good cow and add to the family income. He could buy a thoroughbred sire to head his dairy herd and thereby take one long step toward better farming. Possibly he would build a silo and double the stock carrying capacity of his farm. He may buy a musical instrument or send his son and daughter to the agricultural college.

Of course the place to begin a thing is at the beginning, but which end of this thing is the beginning, is the question that bothers us. In the farm club we begin at the end of "better doing of things" on the farm, hoping thereby to increase the net income so that we may actually do a little real living as we go along. In the Equity Union we begin at the end of "better selling of things," intending thereby to conserve the earnings of the year so that we and our families may "do a little real living as we go along." You see we are both headed for the same goal, though facing in opposite directions.

When the first transcontinental railway was ionstructed in America the building was carried on at both ends at the same time. The two forces of workmen were seeking the same goal, and when the crew from the east and the crew from the west met at Ogden and co-operatively drove the golden spike that completed the great work, then was there great rejoicing on either side. So mote it be.

H. W. WRIGHT, Mott, N. Dak

BETTER CONDITIONS ON THE FARMS

Who is interested in this proposition? We must all be fed from the land. Mother Earth feeds us all, but she requires the right treatment, before she will give forth her rich stores upon which our one hundred million people depend for food and clothing every year. Two thirds of the people of our great country, depend for food and clothing almost entirely upon the other one third, who live on our millions of farms. The farmers must succeed or the people will suffer for the necessities of life.

Professional men, business men, manufaccturers, wholesalers and retailers are awakening to the fact that there is no prosperity on the farms. How to get better conditions is the problem to be solved. We are sure that the appropiation of a million dollars by a machine trust, and another million by a meat trust will not be sufficient to solve this question of right conditions on our farms.

We believe in scientific farming. We are glad our farmers are receiving a lot of instruction along this line. We would not have less education in this direction but more of it. We are glad that our schools and colleges are taking up agricultural studies. The farmers as a class are sadly in need of knowledge in the conservation of soil, the rotation of crops, the handling of stock and dairy products. But there is still a missing link. The chain of success is not complete. The two parts must be linked together. Golden rule co-operation and scientific production must be linked together by organization. The farmer must be "shown" how to defend himself against the business methods of the very people who are putting up millions of dollars to teach him how to produce more for them to take.

The Equity Union has the idea that farmers must be taught better methods of buying and selling before we will have better conditions on our farms. The farmers themselves must make better conditions on the farms by Equity Union golden rule co-operation.

Instead of one million dollars given by a trust to teach better farming, we need to organize in the Equity Union and put the trust out of business, and instead of one million dollars for better farming we will have a hundred million dollars. If you want plenty of satisfied farmers out on the farm, organize them strongly into the Equity Union and give them all the wealth they produce, and they will make this the most prosperous and happy nation in the world. What we need as farmers is to be sure of a fair reward for our labor. We must "get together" in the Equity Union and run our own business co-operatively, more and more every year.

Give us sure prices, not high prices, and reduce the high rents, high interest and high cost of coal and farm machinery and we will make better conditions ourselves on every farm. Co-operative mines, mills and markets are absolutely necessary to better conditions on our farms. Brother member, go to that Equity Union meeting October 3rd, and work for the organization that will make your business more prosperous. You and your neighbors can stand a lot more prosperity, and this will bring it if we get together and stick together.

MR. POOR MAN MEETS THREE CAPITALISTS IN CHICAGO

Mr. P. M. Good morning, gentlemen.

Mr. Caps. Hello, where did you come from?

P. M. I am from northwest Nezraska. I live on a ranch.

Mr. Caps. Have you a family?

P. M. Yes a big one. Wife and I have ten children.

Mr. Caps. Come to Chicago to see the sights?

Mr. P. M. No I brought a couple of car loads of cattle and I am looking for a place to buy coal for next winter. We have a club of 100 farmers at our town and we want good coal at a more reasonable price.

First Cap. Well I own a coal mine but I sell through this jobber. He takes my entire output. I only get 25 cents a ton for my coal clear of expense.

Second Cap: I am a jobber. I only sell to regular dealers. I could not sell to your club. I only make 50 cents a ton clear. I have to send out traveling men to sell and collect. Competition is keen. It cost me a lot of money to get the trade and hold it.

Third Cap. I retail coal out in Nebraska.

Mr. P. M. Oh yes, I know you. Your profit was $2 a ton last winter when it was so cold. Don't you remember? I drove up to your office last winter and I offered you a mortgage on a fine team of horses if you would let me have ten tons of coal and you wouldn't do it. I had to get two neighbors on the note and you charged me ten per cent interest. Now our farmers' club wants me to see if we can't buy direct from a good mine.

Mr. Caps. (In unison.) **We don't sell that way!**

Mr. Equity happens along. Say, Mister, you go down to Greenville, Ill., and enquire for the Equity Union office. **They will sell you coal right.**

The Next Morning.

Mr. Poor man in Equity Union office.

Mr. Poor Man. Good morning, sir!

Mr. Equity. Good morning, come in!

Mr. P. M. Why you have a big priting office here! What paper do you print?

Equity. The Equity Union Exchange, a co-operative paper for the Equity Union.

P. M. That's a big press.

Equity. We need a big one. We print 50,000 copies every week and 16 pages to each paper.

P. M. You don't say! Now I came down here to see about coal. Our club out in Nebraska wants to buy 1000 tons together.

Equity. Well, you fellows take stock in our Equity Union coal mine.

P. M. Have you a mine?

Equity. Yes, we have twenty thousand stockholders among our members. A lot of them are in Nebraska on the Burlington railroad. Our mine is on the Burlington.

Mr. P. M. How much are the shares?

Equity. They are five dollars a share and the limit is two shares. If you take one share and patronize it, we will give you one more for your patronage.

Mr. P. M. Why, twenty thousand shareholders at ten dollars each, makes two hundred thousand dollers. What do you do with all of that money?

Equity. We sunk the shaft and put up all of the machinery. It takes money. Our mine is well equipped. We clean our coal so that our members get the very best six inch lump coal that can be made.

Mr. P. M. How do you sell it?

Equity. Every one of our stockholders get their supply from our mine and they sell all they can to outsiders. We pay the land owner ten cents a ton for the lump and egg coal and we pay the miners for their labor. We save the jobbers profit and expense and the awful hold up of the retailer.

Mr. P. M. Well that looks good to me. I am going home and get every one of my club to join the Equity Union and to take stock in your mine.

Reader, this is a dream, but the writer is working over time to make it real. Equity Union will make its members mill owners, mine owners, and factory owners and save them millions of dollars annually. We are marching on to victory for the common people. **Work with us all you can!**

EQUITY UNION FARMERS STICK

Most farmers would change my caption to a question, "will they stick?" but in the Equity Union we can say "they stick." Every one admits the wonderful power and protection of one million farmers united. No one doubts for one moment the great benefit of organization. We are all sure that if union and co-operation are good for manufacturers, railroads, bankers and all business men as well as the millions of wage earners, it will prove equally as beneficial to the millions of farmers. The soil workers feed and clothe the world. Let farmers stop marketing a short time, and the people cry out with hunger. Our powers of protection and defense are almost unlimited when we act together, from the fact that we hold in our hands once each year the food supply of all other classes. We often hear the declaration, "the farmers could rule the world if they would stick together," "but they won't stick."

This is the knotty problem the Farmers' Equity Union is solving. If you want farmers to stick show them the benefits at their home market in cash. This will hold them together. Our plan makes $ $ $ stick right out. We are able to count out cash right under their noses. This bids for patronage and unites more and more trade. The more trade we get the more economy in handling.

We never pay anything back to non-members for patronage. This brings them in. We say to our unions: "Always hold the door open for all farmers to come in, but never, no never, pay any of the earnings back to non-members." As long as they can get the benefits on the outside they will stay out. Our great purpose is national co-operation. There is a hundred times more in national co-operation than in local co-operation.

One local Exchange can make five or ten thousand dollars annually for two or three hundred farmers if we can get them to unite their trade on our plan. Through national co-operation of a large number of Exchanges, low prices on our largest crops of finest quality can be prevented. While three or four cents per bushel could be made locally, often twenty or thirty cents per bushel could be made by national co-operation, which holds the surplus in the country and prevents a large visible supply. The low price is often caused by too much on the central market at once. This is plainly shown in marketing the 1911 wheat crop when the price was held under the dollar a bushel by putting seventy million bushels on the market at once.

A National Union of farmers will keep each member posted as to the supply and demand and price. When we buy nationally as a unit, we will reduce farm machinery 50 per cent. We will buy the entire output of mines, mills and factories. Farmers, let us be broad minded and build a national union of farmers with a million members. We can do it if we stick to Equity Union principles.

One of the strong points which holds farmers is the fact that each gets $100 invested. He finds that if he only puts in $25 he finally owns four shares or one hundred dollars and then he gets back in cash the twenty-five dollars he put in at first, so that he has one hundred dollars invested which did not cost him one cent. By loaning the union twenty-five dollars he gets it all back and owns one hundred dollars stock in the company, which gives him back from twenty-five to one hundred dollars for his patronage every year, which Mr. Profit Taker always appropriated for his own private use before the farmer joined the union.

We are a new organization and have spent every dollar the first year in planting and cultivating unions. The work of education has been done in a number of communities in Indiana, Illinois, Missouri, Kansas, Nebraska, South and North Dakota and Colorado.

1. Agitate, advocate and educate.
2. Do business at home which shows benefits.
3. Follow golden rule co-operation.
4. Hire a good manager and put him under a good bond. Balance his books every Saturday night.
5. Pay back in cash or shares all you can for patronage. Patronage makes the earnings.
6. Never pay outsiders back anything but let them come in.
7. National co-operation is our goal. Let us all support the National Union.

These seven points will make farmers stick.

A NATIONAL UNION

Brothers, we must keep in mind that we are building a great national union of men, women and children. We will have five million members in the Equity Union family for we count the women and children of every Equity Union family as members. They are a part of this great movement for economic freedom and we want them to feel that they have a part and a responsibility in the Equity Union.

We are building this Union on the idea of the Fatherhood of God and brotherhood of man. On the golden rule principle of love for God and your fellowmen. Of charity for all and malice toward none. Of equal rights and opportunities to all, and special privileges to none. We believe in Christian Socialism. We believe "the people" have the Divine right to rule both in politics and business and we are sure they have the Divine Power to rule when UNITED.

The "Special Interests" do not want us united and so we find hundreds of little "Special Interests" fighting the Equity Union. They don't want the people united and they throw mud at the leaders and yell frantically at the people, "You can't do it" or "You'll hurt business."

As the little Giant Equity Union grows and gets a hold on the business of the country the "Big Special Interests" will finally take notice and then will come our hardest fight. We must hasten to educate the people till they cannot be bought by any special, either large or small. The people must be made to understand that golden rule co-operation is their only hope if they want industrial freedom. We must agitate and educate till "the people" will demand and get banks in the post offices that will loan money at 3 per cent. This will reach the root of our difficulty. It will break the power of the Money Kings. Brothers, why pay ten and fifteen per cent as some of our Equity members are doing. "We the People" have the right to run our own banks and loan ourselves money at 3 per cent. This will be a long step toward economic freedom. We must have Central Equity Union creameries and milk condenseries, Central Equity Union flour mills and packing plants.

We must center our trade together for windmills, silos, gasoline engines, wagons, and all farm machinery and if driven to it we will manufacture our own machinery. All of these things can be done and every member will share in the great benefits. But it will take organization, education and golden rule co-operation. Every new member brings in a whole family and the work of education begins in that home.

A National Union is built of individual families. The families of whole communities are uniting without regard to political belief or religious creed. They are uniting in the Equity Union for the purpose of driving the "special interests" out of their home markets. They are tired of giving the profit takers the wealth they produce, and then when they get hard up on account of the robbery of the system, loan it back to them at 10 per cent interest.

Equity Union will break up this endless chain of robbery and make a chain of co-operative creameries, mills, packing houses, elevators and stores that will establish golden rule co-operation. Brother, go after another member. Get one more shareholder in that Exchange. Get your Exchange to take an ad in our paper.

EQUITY UNION NOT IN RESTRAINT OF TRADE

An effort has been made to show that co-operation by farmers has the same form and effect as an illegal trust in restraint of trade, but there is nothing in this argument. The Equity Union farmer wants direct access to the consumer when he sells, and direct access to the factory when he buys. He is opposed to trust domination and its co-operation because it is co-operation of the few to the sorrow of the many.

We believe in golden rule co-operation because it encourages production and increases trade instead of restraining it. We ought to combine and co-operate against low, unjust prices for farm produce and high prices for farm machinery and all manufactured goods.

The Equity Union can only get full control of the marketing of all farm produce by organizing the consumers WITH US, and breaking down entirely the organized power between producers and consumers, which continually discourages production by low prices, and limits consumption by high prices. This is in restraint of trade and can be checked to some extent by law, but must be entirely eradicated by an industrial Union like the Equity, which finally brings producer and consumer together in the same organization and divides the unnecessary profits between them. The ten cents a dozen difference in price on eggs can be divided b tween producrs and consumers.

The big difference between hogs on hoof and pork on the block can be made an attractive patronage dividend which will hold producers and consumers together in the same union. Equity Union flour mills are coming, not in restraint of trade, but to break the power of the milling trust and give producers better prices for grain and consumers lower prices for flour. That 20 cents a bushel between wheat and flour can be largely eliminated. The 25 to 50 per cent PROFIT on condensed milk, which robs both producer and consumer, SHAMEFULLY, will vanish into thin air when Equity Union grows strong.

When Equity Union has half a million members you will not see a million bushels of good apples rot in eastern orchards, while western people pay seven cents a pound for apples. When farmers unite in our Equity Union and go DIRECT to one reliable factory and buy ten thousand self-binders, headers and mowers together, EACH of the ten thousand members will buy his machine at the same rate as a Capitalist could buy ten thousand. An organization that pays full value for farm machinery and at the same time lowers it to the farmers 50 per cent below the Trust price, is surely not in restraint of trade.

Farmers must get together in the Equity Union and make their business the best in our country by golden rule co-operation.

NO CHARITY NEEDED

The Kansas City Star says: "Charity from the people of Kansas City never can make the farmers of Jackson County prosperous. No class of men ever was made successful by coddling. Alms do not build character. The man who hoes his own row is the man who wins.

Here is a truth that we as farmers ought to lay to heart. God helps those who help themselves. Only those who strive for victory will win. He who would be free must be taught and persuaded to strike the blow himself. Nothing else builds character like strife and struggle against adverse conditions and circumstances. Our very exertions against bad conditions strengthen character and develop manhood. Defeat is often turned into victory and adversity into blessings by farmers who have learned self-reliance by meeting bravely every condition in life.

A Humiliating Position.

However, we would not be misunderstood, and I would not have our farmers misunderstand the appropriations recently made to "show us" how to farm better. I am afraid some farmers accept the millions given by large mail order houses, implement manufacturers, bankers' associations and railroad companies to teach us to produce crops, as magnificent gifts, as charities from a superior and more fortunate class to those who live in a lower sphere. Nothing could be more humiliating and degrading than a position like that by us as farmers.

Cold-Blooded Business Proposition.

These millions of dollars by these millionaire corporations are appropriations made on cold-blooded business principles. They are investing money in the farmers' business. They know that their prosperity depends largely upon agriculture and the success of the farmers. They are so thoroughly organized that they are sure of their share of the prosperity made by the farmers and as long as the farmers are a great unorganized body they know they can take a large part of the farmers' share. The railroads know that the larger the crops the more they get for hauling. But farmers take less money for large crops of fine quality than they receive for small inferior crops. This is caused by want of organization. We compete with each other in marketing large crops and break down our prices below cost of production. When thoroughly organized we will not only produce scientifically but market co-operatively.

Through organization the railroads not only get their share of the prosperity made by the millions of farmers but they take a large part of ours and hand back a million or two each year to teach us how to raise bigger crops for them to haul. The great need of the American farmers today is not charity but organization.

We need an organization like the Farmers' Equity Union which promotes the intelligence, morality and fraternalism of its members and makes them golden rule co-operators. Charity degrades our manhood. Organization and co-operation will render us independent and self-reliant. We never can become permanently successful as a class by depending on others. The old lark taught the farmer to help himself. But the farmers are a little slow to learn that the best way to do this is by helping others.

The golden rule spirit must be promoted among our members. He that would help himself must help others. The beauty of Equity Union is that we cannot help ourselves without helping others.

Equity Union principles make war on selfishness and narrowness. They promote generosity and public spiritedness. The little narrow selfish fellow will never make a co-operator unless we can broaden him out. Farmers, let us work hard for Equity Union and build an organization so strong that we will have protection against every man and every system that exploits us in any way. Through this Union we can make all the wealth we produce come out into the country the grandest place in which to rear our families.

Let us not look to others nor to charity for success, but to fraternalism and co-operation through the Equity Union.

Wm. L. Hearron, Palmyra, Ill. is a pioneer worker for farmers' organizations. He was one of the three charter members of the Farmers' Esuity Union.

LOCAL FARMERS' ELEVATORS NOT SUFFICIENT

Many Local Farmers' elevators are a partial success. They do good. But an Equity Exchange is much better than a Local Farmer's Elevator or a neighborhood club.

The Equity Exchange is organized by a National Union, with prestige, power and influence back of it. The work is done thoroughly and intelligently by an expert organizer, who has clearly in his mind an up-to-date plan of co-operation, which he educates one hundred or more farmers to follow before they start in business. This education of each community where our Exchanges are started goes right on from month to month and year after year. Meetings are held, lectures delivered and a weekly co-operative paper reaches each member weekly.

The National Union is an organizing, educating force, and is worth ten times what it costs. The short sighted, narrow, ignorant man sees no benefit in this campaign of education, because he is the very man who needs it the most. The Equity Exchange which takes out $1 of the fifty or one hundred dollars which the Equity Union has made for this man and pays his national dues has done him a favor as well as every member in the United States.

Through a National Union campaign of education every Exchange is made stronger every year and the dollar dues well spent. The paper is well worth the money.. Through our National Head our members and Exchanges find more and more opportunities for co-operation. This advantage will be more apparent from year to year as we grow stronger and better acquainted.. Some are continually prophesying failure on the farm machinery question, but we are sure of success. The International Harvester Company will run after our business in five years. We are told we "can't" run a mill, but we will run a mill successfully. Let every member support the National Head with eight ane one third cents per month and we will work out successfully every proposition we have undertaken, and many more besides. The farmer who will hold back eight and one third cents from an organization like the Equity Union and spend ten times that amount foolishly each month must be a hay seed in deed and in truth.

Let every Equity Exchange pay the National dues for every member and then send a delegate to our National Meeting in Kansas City, Kans., Grund Hotel, December 16th, 9:00 a. m., and demand a printed financial report showing every dollar the National Union received and expended in 1914. Members must support the National Union and control it. It is their Union.

BUILDING A BUSINESS

We often hear it said, "He has made a fortune in this town in the last fifteen years; he has built up a fine business since he came here." But the question arises, did he build the business of his patrons. He is entitled to good pay for good honest, efficient service. No fair minded man will deny him that. But the most of the fortunes are made by taking large unjust profits from the patrons of a business.

The farmers and many other wealth producers are building up all kinds of businesses for unnecessary middle men throughout the United States by GIVING them their patronage and allowing them to collect unnecessary profits. The middlemen are not to blame for this, nearly so much as the farmers who refuse to unite and SUPPORT the Equity Union its honest sincere efforts to overthrow the profit system.

Wherever it is fully carried out the farmers are building up a business for themselves instead of for unnecessary middlemen. They are making the wealth come out into the country where it is produced instead of building up innumerable businesses for other people.

At over one hundred good markets, the Equity Union is educating by actual DEMONSTRATION, a large community of farmers and their families, that it pays to unite in the Equity Union and build up a co-operative business instead of building a fortune for others. This plan gives the wealth producer all he produces and makes him a satisfied citizen. Equity Union is endeavoring to build up the country and make it a better place in which to live. When this is done the towns and cities will take care of themselves. Every town which is surrounded by prosperous farmers will have all of the prosperity it deserves. We want every Equity Union member to build up your own Equity Exchange. Haul every bushel of your grain to the Equity Exchange. Buy all of your flour, feed, twine, cement, salt and coal together. They are furnished to members at wholesale price in car load lots. You pay the retail price, but all retail profit is paid back to you so that you are really buying at wholesale price in car load lots.

We want our members and Exchanges to buy their farm machinery direct from factories. As long as we support the old system we will keep the great army of traveling men going at our expense and keep on building up colossal fortunes for such millionaire trusts as the International Harvesting Company.

The farmers need the Equity Union at every good market so they will build a business of their own. We rejoice when we see this Union growing because we know it is helping the class of people who have long been making prosperity for other people instead of themselves. We are glad as we think of the true-blue-golden-rule co-operators in the thousands of Equity Union homes, who are determined to quit patronizing grain firms who are organized so thoroughly to sap the country of most of its wealth and center it in the hands of a few rich people.

Brother, remember that every dollars worth of patronage you give a grain company builds their business at the expense of yours, and makes that company more powerful to rob you and your family in the future. Let us build Equity Union a million strong. IT WILL PAY.

NECESSITY OF ORGANIZATION

It is safe to assume that every person believes in organization for almost any purpose, and it is also safe to assume that but few people ever stop to think how necessary is organization to accomplish a purpose. A careful hold-up outfit organizes and plans its work so that every one knows his part of the game. Success would be impossible without it. The successful manufacturer, business concern or establishment of every kind is the one that is well organized.

We have heard some persons treat organizations very lightly. Some people censure a church organization because of some of its members, and they concede to themselves the ability to live the right kind of a life without belonging to any church. That is possible, but how many do it? Some persons say that they can teach their children just as much at home as can be learned at school, if the teacher doesn't suit their fancy. That is also possible, but how many do it? In any realm of life it is absolutely necessary to organize.

I recall one place in this state where a farmers' institute was tried out without any organization back of it, and while in general, the people seemed to think it would be a good thing, yet when the Saturday afternoon for the meeting came, less than a dozen men went to the hall, although there were fifty-two country vehicles on the streets. At night hardly any one attended. The great reason, or the moral to the whole matter, was that no organization was back of it, and only two or three persons seemed to be directly interested. It is a well known fact that a rousing farmers' institute can be held at that place if a strong organization is back of it.

I remember the first institute that was held in a certain town last February. There was no institute organization, but the ethodist minister and his congregation arranged for the institute, and boosted, and they had fine meetings. Farmers complain about the other fellows getting too much of the pudding, but you don't hear the other fellows complaining about the farmers. The other fellows organize and remain organized. Doctors travel all the way across the state to attend a meeting of their profession, but in general the farmers don't want to go across the street to hear an address. Merchants, bankers, lawyers, doctors and dentists all have their organizations and they make use of them. They get results through organization, both directly and indirectly. If an organization is necessary, or even more so, for farmers? Organization promotes sociability, encourages education along the lines for which the organization was promoted and gets results by producing efficiency.

Our breeders of pure-bred stock are organized, and there are many kindred organizations such as the bookkeepers' association, the corn improvers association, horticultural society, and others. Our most progressive farmers belong to some of them but the rank and file of farmers who need the uplift of these organizations the most are too often absent. What could a political party do if it was not thoroughly organized? What can a religious movement expect to accomplish without thorough organization? Nothing.

As a striking example of broken organization, I might mention the lack of spiritual progress in America which is due to the lack of unity of organization. Likewise, one great reason why the farmers do not face the world in a more solid rank is the lack of organization. If the farmers would organize into one great organization for the general uplift of their people, financially, socially and educationally, and remain organized as well as the professions or things to suit themselves.

It might be well to add that it is not advisable to get too many organizations in one community. If the Grange is there, make it the organization of that community; if the Union is there, make it the organization. Let everybody boost. The idea is that if too many organizations are tried in one place, none will succeed as it should. It is like having several little churches in a small place or several lodges in a small place. We all know the resuts.—Nebraska Farmer by James Pearson.

AN HUMBLE BEGINNING RESULTS IN A MIGHTY FORCE FOR UPLIFT OF HUMANITY

Nineteen hundred years ago the "Despised Nazareen" started the Christian religion with twelve poor fisherman as his apostles. Ignorance, selfishness, superstition and degradation reigned supreme even among his own people, the only nation which had a knowledge of the true God. The other nations were in midnight darkness of heathendom. The Nazareen said "Ye shall know the truth and the truth shall make you free." This was the slogan of his disciples as they spread thru the world and built up the great cause of Christianity.

This organization started by the "Despised Nazareen" and twelve poor fisherman has encircled the world with its blessings to humanity and all other reform movements have or will come as a result of Christianity. As humanity becomes more intelligent, moral and fraternal, they will adopt and follow golden rule co-operation.

Rochdale System

About sixty years ago the principles of co-operation were formulated by poor cotton weavers in England. They were as despised as the Nazareen himself. But their principles were pure and humane. 1. They said, "That human society is a brotherhood, not a collection of warring atoms. 2. That true workers should be fellow-workers, not rivals. 3. That a principle of justice and not of selfishness should regulate exchanges. Mark the growth of the Rochdale System from a few poor despised cotton weaver. In England and Scotland alone there are more than two and one half million stock holders in more than fifteen hundred retail co-operative societies which together own the Co-operative Wholesale Society of Manchester, England and the Scottish Co-operative Wholesale Society of Glasgow, Scotland.

The families and friends affiliated with these stockholders include nearly one-fourth of the entire population of Great Britian. The cost of living is materially reduced and the character of the co-operators is greatly improved. Let us never despise a movement because of its small beginnings. Evil is the only thing that leaps full-grown into the arena of combat. Sin leaped into our world with more than man's strength. But thru Christianity and co-operation every enemy of humanity will be overthrown.

Equity Union No Exception.

Three plow-handle farmers obtained a national charter from the Secretary of State, Springfield, Ill., duly sealed by the great State of Illinois and signed by the Secretary of State, Dec. 16, 1911. Under this charter we can charter a local union anywhere in the United States. The only asset we had was a legal charter, a well defined plan of co-operation and one man who had so much faith in the proposition that he was willing to spend his own money and time in traveling, advertising, holding meetings in Indiana, Illinois, Missouri, Kansas, Oklahoma, Minnesota, North Dakota and South Dakota. Being able to leave behind in the home of every new member a weekly paper which teaches true blue golden rule co-operation has been a most potent force for good in the development of our Equity Union.

The campaign of organization and education has been incessant for four years, both in winter and summer. The actual demonstration of golden rule co-operation in twelve states, where we have started the Equity Union is educating the people rapidly and we hope to widen the twelve circles so well begun so that our membership, influence and power for good will double every year and our Union will be stronger than the Rochdale Societies of Europe.

It is founded on right principles? Is the plan of co-operation practical and comprehensive? Is it being pushed vigorously? Then it will succeed. Despise not the day of small beginnings.

THE EQUITY UNION SELF-BINDER A FIFTY-YEAR REVIEW

I remember when I was five years old, seeing my father cut wheat with a cradle. Then we had a reaper pulled by two horses, but a man raked off. I remember walking two miles when ten years old to see a reaper with a selfrake. For a number of years I bound wheat after a McCormick selfrake, which went over a sweep as the reel revolved and raked off the enormous bundles of heavy Walker wheat in 1860 to 1870.

Then the Marsh Bros. gave us the Marsh harvester, which elevated the wheat over the side for two men to bind as they rode on the machine. After this idea was developed, it did not take long to replace the men with a wire binder, and finally a string or twine binder, and today, after fifty years, we have come up from the cradle as a harvesting machine to the modern self-binder which works almost to perfection, and enables us to harvest more than one hundred million acres of grain every year.

Progress was made slowly but surely, step by step. Many mistakes were made, but they were only used as stepping stones in educational process. When we used the first reapers there were inventive dreamers who looked forward to the time when we would throw off the bundles of golden grain on the side, well bound by a self-binder; but the pessimistic stand-patters said it was impossible—never had been done and never would be. But we are doing it, and people have ceased to wonder at the wonderful self-binder.

The Farmers' Problems.

As the inventors and manufacturers have wrestled with the problem of harvesting and threshing to feed the one hundred million people of the United States, so the farmers, for 50 years, have struggled with the problem of uniting in a strong, fraternal national union that would enable them to present a solid front, a wall of defense, against the trusts and combinations which have grown up to exploit them on every side. The Grange was very strong in Southern Illinois when I was a boy, but its ideas of co-operation were very crude. It disappeared with the wheat cradle and man-rake reaper.

The F. M. B. A. was two hundred thousand strong in 1890, and the alliance had nearly one million members and showed many improvements over former organizations. But these unions had not enough of the elements of true blue golden rule co-operation to make them adhesive. Instead of developing these important principles among the farmers, they allowed politicians to creep in and break them up. These organizations were not a success, but educators. They were stepping stones to something better.

The American Society of Equity spread over a number of northern states very rapidly in 1902-1907, but it united any and all classes of people who would become subscribers to a certain newspaper and was a rope of sand. The experience some of us received in the A. S. of E. was very beneficial in starting a new union and in building it up successfully.

Not a Branch.

The Farmers' Equity Union is not a branch of the defunct A. S. of E. It has no connection and never has had. But the founders of this growing Union, which has made a good start in twelve states, are building on the foundation and failures of all past farmers' organizations of the last fifty years. We are profiting by fifty years' experience. We are cutting out the weaknesses and handicaps of the old unions. We have only one head, a National Head. Instead of state unions, every state is represented on the national board of directors. There is no horde of county and state officers to absorb the revenue. Every dollar of revenue goes to some one who is showing results in building up the Union and educating the farmers to be golden rule co-operators.

The Plan of Co-operation.

The Equity Union has a business plan of co-operation that unites the farmers and keeps them united. It enables them to run a safe, prosperous business that shows dollars and cents every year for its members. One farmer gave me a letter which he had received from an Equity Union member who had a taste of our co-operation. This is what he said: "You ask me what I think of the Equity System. I think it the greatest system ever gotten up to unite the farmers—to make them co-operators—to return to the farmers the profits on their patronage—to unite the farmers so they can buy at wholesale prices, etc."

This farmer has "tried out" the Farmers' Equity Union in handling a very large wheat crop. If you ask him his opinion he will not give you a theory, but the facts. He will tell you that the Equity Union is no wheat cradle, man-rake, self-rake or dropper, but a genuine self-binder. It binds the farmers together as no other Union is doing or ever has done.

EQUITY UNION A REMEDY

Every one who has studied Equity Union principles is convinced that through this organization great reforms will be worked out in the economic, social and political world. When a new Union is organized they are set to the task of saving money for the members at once, through co-operation. Three or four hundred dollars is saved on a car load of twine by buying in car load lots direct from the factory. The supply of coal for the entire Union is bought direct from the mine in car load lots. This move is an educator and sometimes quite an eye opener. It gives the people a taste for more co-operation. They will meet together and combine their orders for a car of salt, flour, fed or cement.

The first lesson they learn is that there is economy in combination for "the people" as well as for the "millionaires." They learn that "whosoever will may come" in the Equity Union and receive economic blessings. The whole movement is economic in its nature and world wide in its reach toward humanity. But Equity Union does not stop with ordering a few car loads of twine, coal or salt. We build the Local Union up to one hundred or more regular members and including the families we have an organization of five or six hundred people.

We do not leave them here. Our next thought is for a business organization, equipped with grain elevator, warehouse and coal shed and capital in the bank. We want a ten thousand dollar Union at every good town to start with. Then the business world will recognize us and respect us. We will be rated by Dunn and Bradstreet. We will be a real business organization and a people's Union at the same time, reaching down in benefits to the humblest member.

The man becomes the unit instead of the dollar. We are defeating the idea of rich men's dollars first, and humanity's rights last. This reform is worth all of our efforts. But we do not propose to stop here. We intend to overthrow the profit-taking system, "for our members," every where in the economic world, and "our members" must include every honest, moral person who works for his living and wishes to be a member.

We work for benefits for our members, and hold the door wide open for "whosoever is honest" to come in. The next move is to unite all of these Equity Exchanges under one national head. Then they will act together nationally and accomplish one hundred times more for each individual than they ever could locally.

If the agricultural press would be broadminded and unselfish enough to urge the farmers to unite nationally under one head, they would do them ten times more good than to educate them to orginize in clubs, but warn them to beware of those who would unite them nationally. Every Equity Union is a farmer's club or can be made such in every sense of the word, but each Equity Union is also a part of a great National movement working for great reforms in the political world.

We are learning and showing in the Equity Union that a great national union of the people can be organized and steadily built up, without grafters or political place hunters to run it. We warn our honest members to be on the look out for these gentlemen, and especially for the little narrow pin heads, who sometimes get on the board of directors of our Equity Exchanges and then turn traitor to the national union and try their best to kick it out.

We rejoice as we think of the splendid leaders in every town who are true-blue for Equity Union both local and national and who will stand by us in carrying out many great reforms through the Equity Union for the common people, God's kings and queens.

AN ORGANIZED MARKET

The most of our leading markets are organized markets. They are organized by and for the few who run them, for their special benefit. The middle men who control them have organized or agreed upon prices, for self protection against cut-throat competition. They all pay about the same price for grain, stock, milk, eggs, poultry and all farm produce. They unite in holding up prices when they sell, whether they are manufacturers, wholesalers or retailers. So that the unorganized or individual farmer has no power whatever to protect himself in such a market. He sells in an organized market and buys in a combined market. His neglect of the most powerful weapon he can find, makes him the slave of all organized classes. He is at their mercy every time he succeeds in producing a bumper crop of any kind. The price-hammerers control very many of our country markets, simply because the farmers are not willing to unite and support an organization. Ignorance, suspicion and selfishness separate them and the organized few control and rob them.

In March 1914, fifty-seven organized hay buyers in Kansas City, Mo., met and advanced the price 50 per cent of handling hay in that big hay market. The farmers held a big indignation meeting and protested against the outrage, but the organized buyers are still in control of that organized market and the farmers are paying their price for handling hay.

Fifty-seven organized buyers control ten thousand unorganized sellers. Five grain companies own five grain elevators at one good grain market and take ten or fifteen thousand dollars unnecessary profits out of every good crop. Five men called capitalists, control the market against three or four hundred farmers who refuse to unite and dig up one dollar per year for an organization, that would give the farmers full control of that market and some years save them fifteen thousand dollars unnecessary profits.

Three men "get together," build a milk condensery and control the markets at one town and make the farmers pay for their entire plant, then go to the second market, erect a fine plant with all modern equipments and make the two bunches of farmers pay for the second plant. Then a third plant is erected and three communities of farmers buckle in, work early and late and skimp and save, until that fine modern condensery is fully paid for. And so on, the "controlled market" is made to do duty, till nine big milk condenseries are paid for and still the unorganized hayseeds own no condenseries. They have simply worked for a bare existence and the "organized market" has made a few men millionaires.

A few men start in the meat-packing business. They erect the most modern packing plant with all improved machinery, they hire and train the most skilled labor at a low price and organize thoroughly with all other packers. This "organized market" controls the price to both producer and consumer, and makes millionaires of a few men who own all the plants for packing meat. Every trust in our country has control of some necessity of life, through thorough organization and acts as a toll gate and catches the people coming and going. The "organized system" is by the few and for the few. The only hope of the producers and consumers is to organize every market and run it in the interest of the common people. Producers and consumers must come together in the Equity Union and destroy the "organized markets" now owned and controlled by the few called capitalists. If you are a farmer or wage earner, join the Equity Union. It will pay.

WON'T STICK

The first and strongest objection to a farmers' Union is "they won't stick." When we tell a farmer or business man that we are organizing a farmers' union, he looks upon us with suspicion at once. The thought seems strong in his mind, that I have a scheme to rob the farmers. Tell him that we are succeeding in the Equity Union in uniting the farmers and he says, "show me."

The idea, that farmers "won't stick" is so prevalent that the most of the farmers honestly believe it, until they are shown. But we take issue with this idea, which stands so in the way of all efforts to unite the great mass of farmers. We affirm that they are the greatest "stickers" on earth and we will prove it.

I have been watching them for 50 years, "stick" to the profit-system like leeches, when they knew they were being robbed to death by it. We have tried for fifty years to pull them loose from a "business system" that takes millions of dollars of the wealth they produce annually and bestows it on the unworthy, idle, extravagant rich. But the more we have pulled the closer some of them "stick" to the old system. I have seen them "stick" to a milk trust and support it faithfully and tenaciously until the trust owned one hundred fine condenseries, all built with the famers' money, which brought them a fine revenue of millions of dollars annually, all the farmers' money, which he and his family worked hard for early and late, and today you could not pull some of them away from the support of that old robber system—No! not with a team of horses!

We have been watching for 50 years the wonderful growth of the beef trust. This trust has accumulated millions of dollars. Brother farmer, if you don't believe it, go to north of Chicago and view the palatial residence and magnificent grounds of a meat packer. Take a look into his private garage. Does he ride in a Ford? No! He owns a dozen of the finest automobiles turned out by the factories. Where and how did he get his money! The farmers "stuck" to the beef trust until they have made that "set," as rich as the meat packers. They have "stayed" with the lumber trust, the machine trust, the sugar, coffee, salt, match and clothing trust until millions of honest, industrious farmers are brought down to a bare existence, after fifty years of hard labor and close economy.

Don't tell me "they won't stick." I am a farmer, and have worked hard in four organizations trying to pull the farmers away from the old rotten system, which constantly made the rich richer and the poor, poorer. And yet today, we go into communities in Illinois and find the farmers nearly all stand patters, slaving on in the same old style, pulling themselves and family to death in the same old rut. Supporting the same old system which has robbed them for fifty years.

The Equity Union has a grand mission. It is doing a work that no other Union is doing. Our mission is to unite the farmers in a successful business organization and keep them united. The education we gave them showed results in 1913. They stuck to their own business and saved one million dollars. By practical demonstration, they are showing their members that it pays to stick to their own business instead of sticking to the grain companies' business.

This is a good lesson to learn. We are showing plainly that what we have been affirming for years is actually true. Farmers will stick if we "show them" that it pays.

Last May and June we held big celebrations with our Unions. They had a fine feast when they had made by sticking together in business for one year. We will divide more melons and bigger and richer melons next year. We warn every Exchange to keep on a safe margin. We urge every member to haul his grain to his own Exchange. It won't pay to support the enemy. We have stuck to that kind of a system too long. We want every member to stick to the Union. Our aim is to build "the little giant" a million strong. We will succeed if every Equity Exchange will pay the annual dues for each stockholder in November of each year.

A million farmers united so they will "stick" to their own business, will in a few years pack their own meat, grind their own wheat, make their own butter and cheese and distribute their food products direct to the consumer. We will break the power of every trust to rob the people in a few decades. The people must be educated to "stick together." It can and must be done. Give us a million families reading our co-operative paper weekly, and we will have a grand industrial union that will stick to golden rule co-operation instead of sticking like leeches to every trust in our country.

OUTRAGEOUS PROFITS PILED ON PROFITS

The Bureau of Corporations presists in bothering the Steel Trust with awkward statistics. Its latest effusion is caluculated to show advantages of combination through holding companies in concealing real costs of production and excessive profits. As a miner of ores for itself, the Steel Trust charges a profit to itself. As a transporter of ore to its furnaces, it charges another profit to itself. As a producer of pig iron, it caluclates another profit. As a producer of steel from the pig iron, it charges another profit based on prior costs, which include all these prior profits.

This is an integration of cumulative profits so well concealed as to make the "book cost" of producing steel rails, for example, some $21.50 a ton, when the actual net cost is only about $16.50 a ton. The Penroses of the Senate are not shrieking for a tariff hearing on this phase of the question. The light is gradually being let in on this infamous system, and as light dispels darkness, publicity will inform the people and show them the nefarious methods by which they are robbed of the immense wealth they produce annually.

These public investigations by the government are eye-openers to the people and will arouse them and educate them to condemn the entire profit-taking system. Every Equity Union worker can rejoice and be glad that so many forces are at work to assist us in breaking down the profit system. The people are being prepared for Equity Union Exchanges very rapidly. The magazines, agricultural papers, and daily press are all educating toward golden rule co-operation. Let every worker take courage and go forward to victory. We are fighting a winning battle. More and more forces are rallying to our side and helping on the good cause.

The mission of the Farmers' Equity Union is to demonstrate true blue co-operation, and bring the blessing in reach of every wealth producer. Every producer of wealth can come in. Only profit-takers capitalists, and grafters are excluded. We pay all profits back to patrons according to patronage. We welcome every one to membership who has patronage for our Equity Exchange. We never pay capital over 5 per cent. This includes Mr. Capitalists. We carry on a continual campaign of organization and education in country school houses, which unites the farmers and centers their trade more and more and reduces the cost of handling.

Our next national meeting will take up the questions of insuring our elevators and bonding our managers through our own mutual Equity companies. Also of owning our own coal mine and buying farm machinery together. These are questions we are determined to tackle and work out for the benefit of our members and as a demonstration for all farmers' organizations. We are shipping millions of bushels of grain through the Equity channel in the Dakotas, to the advantage of our members, and we are working for the same benefits for our Kansas members. Our members in the Dakotas are working for the big June Equity rallies being held in those states by the National Union. Every member in the United States has reason to be encouraged as never before. There are hundreds of forces at work. There are hundreds of forces at work in our country educating the people away from the profit system to golden rule co-operation. The Equity Union is only one of the mighty forces at work for the economic freedom of the millions of farmers and wage earners. It is our mission to demonstrate in as many markets as possible true blue, golden rule co-operation.

HOW TO INCREASE PRODUCTION

Strenous efforts are being made to teach farmers how to produce more. Millions of dollars are being spent by the bankers, machine trusts, Sears & Roebuck, railroads, and now the meat packers have voted half a million dollars from their loot to instruct cattle raisers how to increase the production. At their banquet the plates were one hundred dollars each!

I believe in intelligent farming. I am sure a successful farmer needs more education than a banker or merchant. But the day has come when co-operative marketing must go hand in hand with scientfic production. Until this is accomplished the farmers will not produce more because our most intelligent boys will not stay on the farm. We rear the healthy, strong boys on our farms, graduate them at the high schools and business colleges for the trusts and combinations to use against us in the business world.

The farmers furnish the brains to run the trusts, every one of which is against them. Our bright, intelligent farmer boys must be induced to stay on the farms or used to run a system of co-operative marketing, manufacturing, mining and railroading that is in the interests of the whole people instead of the few millionaires who now own trusts. Farmers, listen. We can much better afford to pay those bright men $200 a month to run a co-operative business for us, than to run the trusts against us. If we raise them, educate them and pay them, we ought to organize them to work for us instead of against us.

BOND COUNTY MILK

The Equity Union is trying hard to unite the Bond County, Ill., farmers into an Equity Exchange and market their milk together direct to St. Louis consumers, whe are paying from three to five dollars a hundred while the farmers' average price in 1913 was less than $1.65 per hundred.

On the Equity Union plan we can raise the farmers' average price one cent a quart, four cents a gallon, fifty cents a hundred or two hundred thousand dollars a year for Bond County milk, and at the same time lower the consumers' price in St. Louis, one cent a quart, four cents a gallon, fifty cents a hundred, or $200,000.

If every milk producer in Bond County will come into the union this can be done. Give the farmers of this little county $200,000 a year more for their milk and they will increase the supply 25 per cent without any more instruction as to how to produce.

Hogs and Cattle.

The seven million farmers of the United States can produce a full supply of hogs and cattle and will produce them when they organize in the Equity Union and market co-operatively. If every farmer in the United States would produce five more fat hogs in 1914 than in 1913 we would sell them below cost of production and half of us quit again the next year. We would crowd the central markets with fat hogs and enrich the meat packers and butchers with low priced hogs, but furnish very little relief to consumers. The farmers of Denmark are getting a living price for fat hogs, and producing not only a full supply for home consumption but for export. They are co-operators.

Wheat Growers.

If the wheat growers could follow the instruction given them on production sufficiently to make our fifty million acres yield two bushels more per acre in 1915 than in 1914, we would have an eight hundred million bushel crop instead of seven hundred million bushels. But how much would the farmers get for it? The railroads would make millions hauling it, the speculators holding it, millers grinding it, and stores retailing it, but millions of farmers would be saying, "It don't pay to raise wheat," and flour and feed would be very little lower. Every intelligent farmer knows these are facts. If he reads this, he knows also that there is a Farmers' Equity Union organizing in twelve states. Wherever the farmers unite on this plan they are succeeding. In 1912 at Liberal, Kan., they saved seven thousand dollars by co-operative marketing. At Mott, N. D., over nine thousand dollars. Every Equity Union is saving money for its members. We keep the other fellow from taking what we make.

The Equity Union puts a weekly paper in every member's home which educates him to market co-operatively. Give him enough of this education and the meat packer, machine trusts and bankers will not need to appropriate millions of dollars to educate us to produce more. They will not have so many millions to appropriate. Encourage production by making the farmers price sure and equitable.

WHO IS BENEFITED?

Geo. F. Baker is chairman of the Board of Directors of the First National Bank of New York City and intimate personal business associate of J. Pierpont Morgan. He told the House investigating committee that since the organization of that bank with a capital of $500,000, it had made more than eighty-millions of dollars profits. This is a fair sample of what is going on in our business world from one end of the country to the other. Enormous fortunes have been amassed in our land simply because the people support the Profit-System.

The New York World says, "National Banks are chartered by the United States government. They have powers and privileges that are denied to all other banks. Among these powers is that of issuing money, which is an attribute itself. Congress could this very day repeal the act under which these banks were created and send every one of them into liquidation. It could by a single law destroys all those privileges under which the First National Bank, for example, has made eighty million dollars in profits on an original investment of $500,000."

The Equity Union is not asking for an equal division of wealth in the United States. We do not want the thrifty people to share their prosperity with shiftless hobos and tramps. We do not believe in condemning men who accumulate wealth by industry and economy. We believe this class deserves commendation instead of condemnation. But we are radically opposed to our government granting special privileges to any class which enables them to rob the masses. This has been going on till the combination of the privileged few now has absolute control of our money system and of very large per cent of our industries. This power can only be wrested from the combinations by the people. The people must be organized into industrial unions and educated to be fraternal co-operators.

A union, with a strong organizing force behind it, which carries on with more and more power, a campaign of organization and education, and which continually promotes the intelligence, morality and fraternalism of its members will finally make the people masters of the situation in politics and business.

Responsibility.

The people are sovereign in America. They have the Divine Right to rule. Upon them rests the responsibility for good government and of equal rights to all and special privileges to none, not only in politics but in business. Under the present system in vogue in our country the millionaires are the masters and the people subservient vassals. The capitalistic class seek more and more power and wealth through combinations of the few to the sorrow of the many.

The people must unite and break down the power of these unholy oligarches by golden rule co-operation. The Farmers' Equity Union is beginning a demonstration in twelve states of a practicle plan of co-operation that will educate the farmers and other wealth producers away from capitalism, from the profit-system and from the big dividend idea. This plan of co-operation makes the people their own capitalists, destroys the profit system and makes big dividends on capital can ever command in an Equity Exchange. No profit whatever is alowed in a true Equity Exchange for handling produce or merchandise. Big dividends are out of the question as every Board of Directors is prohibited from declaring over 5 per cent dividends on the stock subscribed. The justice of true co-operation appeals to the people and they are rallying around the Equity Union banner as fast as they understand our plan and principles.

The National Union is young, but as strong in good principles as Gibraltar, and as sure to win as the triumph of right is certain. The profit-system gives its benefits to the few unworthy. Co-operation benefits the many worthy ones. Big dividends on capital enables a few parasites to filch the wealth produced by the millions of laboring people. Co-operation blesses the common people. It brings them near together, makes them friendly, and gives them power over every enemy. The farmer who refuses to be a co-oprator stands in his own light and hinders one of the best causes ever started in America. We want from one to three hundred co-operators united at every good market. They must have from ten to thirty thousand dollars capital. These Exchanges must buy coal, twine, fencing, fence posts, fertilizer, flour, feed and farm machinery together. They must buy and sell together on a national scale. The campaign of organization and education must be kept going winter and summer.

Our entire business system must be changed gradually, but surely, until the wealth producers and their real helpers are masters in the economic world, instead of selfish millionaires whose greed grows stronger and stronger as we heap millions into their unholy coffers. We want every person to have equal rights in politics and equal opportunities in business. This will come through industrial unions which teach and demonstrate golden rule co-operation. It will encourage every worker and bring peace and prosperity to our ninety million people.

"We are now paying a little over fifteen dollars per hundred weight. Other private dealers pay the same price, but with the private dealer the first return is all the farmer will ever get for his pigs. Each year we lay aside something to pay off what we borrowed at the bank. After we have done this we declare an annual dividend, which amounts to from $1.25 to $1.50 on each hog that has been delivered to us. You see also that at the end of the ten years' period each farmer has a share in our establishment here, which may be termed an additional price for his pigs. Under our system each farmer has an interest in this concern, when it is finally paid for, in proportion to the number pounds of pork which he has delivered to us during all these years. We figure that the average amount, which has been laid aside and invested in this plant, is about twenty-five dollars per member. We do not pay this in cash to them, but issue a certificate which is evidence that they own a share in this plant and in the business which we are doing here. This is not a closed corporation, but any farmer who wishes to join it can do so by paying the estimate value of membership.

Standing Together.

"There are 43 co-operative bacon factories in Denmark. We have a central organization, which is rather a voluntary association for the mutual benefit of the various co-operative bacon factories. The office is in Copenhagen. Weekly reports come in from each factory giving the amount killed and sold, the expense of the business and the market returns received. The heads of the various factories meet from time to time to talk over the best business methods and possible improvements in our way of handling bacon. We give each other the benefit of our experience and think help of every way possible to help each other. We do not feel that we are, in any antagonistic sense, rivals. We fully believe that every factory is helped by the success of the other factories. The success of each depends upon the fact that all of the factories are putting out a good product and are dealing in an honest business-like way with the foreign retailers who take our products. What hurts one of us hurts all of us. We are anxious, therefore, to help each other in every way, since in helping others we are helping ourselves.

"Our agricultural schools and our government departments help us, particularly upon all scientific problems. They help us along the technical side of all our work. They make experiments and give advice and co-operate with us generally in a thousand ways."

Possibly the most striking thing about the factory is that a group of farmers should run a concern that rivals in efficiency and business methods the largest and best privately owned packing houses of the world. We expect farmers to farm well but we do not expect them to do business well. In America they take what is given them for the raw product and go no further. Here they go so far as to get all there is in it. The farmer who raises the pig holds to it and keeps it as his property until it lands in the retail shops of England. All intervening profits are his own. Denmark presents to the world the scientific farmer who is an efficient business man. Will the American farmer ever attain that position?—The Republic, Pratt, Kansas.

SPREAD OF CO-OPERATION

I am a firm believer in co-operation. I am glad to see the underlying principle gaining such headway in this country. A great many people have the wrong idea; they seem to think that co-operation is some new and passing fad. Co-operation is nothing more nor less than the working out, in practice, of the old, old theory that two heads are better than one.

I believe that co-operation among farmers is one of the best things that can be brought about; it unites them in a way which nothing else has ever succeeded in doing. They may organize co-operative creameries, fruit exchanges, grain elevators—whatever they will—if they will only live up to their articles of incorporation and be loyal to the organization they are almost sure to come out winners in the long run. The individual farmer is all too often at the mercy of the man who buys. He is obliged to accept any price that is offered. One farmer, acting by himself, can do little or nothing to change this condition. But fifty or a hundred farmers, acting together as a unit, can do much to bring about such a change of conditions as will benefit them financially and socially. There must be good and sufficient reason, indeed, when one state alone (Wisconsin) organizes over 1,500 farmers' co-operative companies. I have seen it stated that no less than 5,000,000 of our rural population are interested, directly or indirectly, in co-operation. Do you begin to see how this movement is sweeping rural America?

I have recently been gathering some facts and figures on this subject and they astound me—the co-operation movement is gaining ground so rapidly. I am going to pass some of these figures on to you. In 1911, I find, there were 1,800 farmers' co-operative grain elevator companies in the United States, distributed as follows:

Iowa	327
North Dakota	315
Minnesota	266
Illinois	235
South Dakota	222
Nebraska	193
Kansas	126
Wisconsin	38
Oklahoma	33
Indiana	24
Michigan	20

Washington	18
Montana	16
Ohio	14
Texas	5
Colorado	4
Oregon	3
Missouri	3
Arkansas	2
Idaho	1
Kentucky	1

Today this number has increased to almost 2,200. The membership in an association averages from seventy to 225. The output of the large elevators usually varies from 40,000 to 100,000 bushels of grain some of the largest handling a million bushels. The smaller elevators holding from 20,000 to 25,000 bushels, cost from $2,600 to $3,000. This means that there are today not fewer than 275,000 farmers connected with the co-operative elevator associations, that their investment approximates $22,000,000, and that they handle 300,000,000 bushels of grain, or 45 per cent of the total amount shipped from the selections where these elevators are located.

But these farmers' elevators do not content themselvs with shipping and selling grain; many of them buy supplies for their members. Thus in Minnesota, with 266 such elevators, the total annual volume of business is $24,000,000 of which $22,000,000 represents the value of grain marketed and $2,000,000 the value of supplies purchased for members. Sixty three per cent of these elevators report buying coal; 41 per cent, feed; 40 per cent, flour; 35 per cent, binding twine; 18 per cent, seeds; and 16 per cent, salt. Among other commodities bought by these elevators and sold to their members at practically cost are cement, tile, farm machinery, lumber, fence posts, oil and wire fencing. Nine-tenths of the total number handle some other commodity than grain.

Aggravated into taking the grain selling business into their own hands, the farmers around Rockwell, Iowa, organized the first co-operative elevator company in 1889. Eleven years later two more companies were formed. In 1904 thirteen had been organized, and in 1911 the number in Iowa had increased to 327 and the movement had spread throughout the entire middle West. Last year the farmers' elevator companies in Iowa handled 65,000,000 bushels of grain and purchased 200,000 tons of coal, $750,000 worth of lumber, machinery, flour, feed and other supplies used by farmers.

Among dairy farmers I find the movement taking equally firm hold, so that today of 6,300 creameries and 3,846 cheese factories in the United States, 2,120 creameries, or 33.6 per cent, and 349 cheese factories were organized and are now being operated by farmers as co-operative institutions, and at last reports these creameries were distributed as follows:

Minnesota	608
Wisconsin	347
Iowa	313
New York	118
Michigan	101
Indiana	77
Illinois	55
Pennsylvania	92
Vermont	59

Of the co-operative cheese factories 244 were in Wisconsin, thirty-nine in New York and twenty-four in Minnesota. These co-operative creameries have resulted not only in taking the hard work of butter making off the overworked farm wife, but in improving and standardizing the butter product of the various communities, with better prices resulting. In many places eggs and poultry are being marketed through the same channels, greatly reducing the former 78 per cent loss as a result of the improper handling of eggs from the farm to the market, and securing better prices because of the product being standarized and marketed under a guaranteed brand and stamp.

Do you begin to see the bigness of my subject? Would so many hundreds of thousands of farmers become enthusiastic converts to co-operation if it were merely a fad, short lived, of doubtful value? You and I will need to study this subject carefully and fully in the coming years. It is a big subject, one of the biggest and most important now confronting us. To it, I am convinced, we must look for the solution of many of those problems which now vex us sorely and which we are unable to solve so long as we fight single handed.—James M. Pearce, in Iowa Homestead.

WHY FARMERS ARE NOT MILLIONAIRES

From much that we read in newspapers and farm journals these days about farming and the wonderful profits to be made therefrom, one could naturally expect to meet frequently with millionaires who had amassed their wealth from the farm, but the writer has never met with any of this brand of millionaires nor does he know of any one who has. It has always happened in other kinds of business, that when they became extremely profitable, people rushed into them attracted by the large profits; but no such rush is noticeable on the farm. On the contrary there is such a Nationwide Exodus from the farm as to be a matter of grave National Concern; but why should such a condition exist? And who is responsible for it?

It has been assumed by some that the farmer was wholly or largely the cause; that he was lazy or incompetent, or shiftless, or all of them; while these things may possibly be true of a small per cent of farmers, they whould be equally true of any other business or profession. So that we must look elsewhere for our solution. Those who have thought that the farmer, like Nebuchadnezzer, was a little bit light weight, have sought to send us farm demonstrators and the like. But let us turn to facts furnished us by the office of farm management at Washington, which shows that the net labor income of the Nebraska farmer and his family is about $142 a year and in Iowa about $199 per year. The labor income is determined by educating from the gross receipts of the farm; the money spent in producing the crop, paying taxes, machinery kept in necessary repairs, and allowing 5 per cent on the assessed value of the farm.

The following I quote from Mr. G. L. Carlson, Editor of Carlson's Rural Review, he says: "I have obtained hundreds of statements from well-to-do farmers as to their condition and progress and have selected this from a farmer in Northeast Nebraska. This statement was made November 30th, 1913. This man paid $45 an acre for well improved quarter section in March 1900. Next March he will have owned his farm 15 years. He is considered as good a farmer as lives in his neighborhood. He paid $5000 cash on his farm at the time of purchase, leaving an unpaid balance of $2200 at six per cent interest at the time he bought this farm and moved on to it. An inventory of his personal property showed a value of $3150. He had $1185 in cash. This gave him a capital of $9335 invested in his business at the time of the purchase of his farm. In the fourteen crops grown on his farm during the fourteen seasons, this man has met with no serious reverses. He has three children and these he has educated farily well which he records as an asset; his family is economical and thrifty; he has never owned an automobile. His farm accounting is a very good record having kept account of every transaction during the 14 years.

The following is his statement:

First capital invested	$9335
Balance paid on farm	2200
Invested in education	915
Improvements made on farm	2365
Inc. in personal property	135
Cash on hand	610
Total	$15,560

Depreciation of farm buildings was placed at $250 which deducted from assets, leaves $15,310. Of this $9335 was original capital, leaving $5975 more in assets than he had at the beginnig. If we consider the capital invested $9335, worth 5 per cent simple interest for the 14 years, the interest would amount to $6534.50 or a loss of 14 years labor and $559.50 besides."

Mr. Carlson gives as his conclusion that, "If farmers of the United States were to be paid a living wage for their labor, the farms would pay two per cent on their present selling price When you compare this with 16 per cent National Bank Dividends, 35 per cent Packers Dividends, and 40 per cent Express Co. Dividends, it may explain to some degree why people are not falling over each other to get on farms.

Then there is the tariff; the farmer is expected to pay a higher price for the clothing he wears in order that the New England manufacturer may get a higher price for his manufactured goods and then the government takes the import duty off from foreign wool and puts the farmer in competition with the wool growers of the world so that the aforesaid New England manufacturer may get wool more cheaply. This government allows the National Banker to buy its interest bearing bonds on which it pays no taxes, and then issues the full amount of the bonds in National Bank notes and loan them out and get another interest. It allows whiskey men to deposit their liquor in Government bonds, warehouses and then issues them a negotiable warehouse receipt which has many of the functions of money. Until recently railways and express companies have charged unreasonable and excessive rates. Secretary Houston has quite recently told us that "four Packers in this country made more profit last year than all the breeders and farmers who produced the livestock."

Up in Minnesota last winter they investigated the Minneapolis Grain Combine and discovered that they were robbing the Northwestern farmers of $14,000,000 annually. This Combine was ably assisted by the banks and daily newspapers.

Then farmers and stockmen want to borrow money in this country. They pay 10 per cent interest whether on bank loans or real estate mortgage loans. These are some of the reasons why millionaire farmers are not in evidence and also why some of us have concluded that we do not care for any more farm demonstrators until we have solved the problems of Co-operative marketing and have built a few Co-operative Packing Plants and have organized a few Farmers' Co-operative Credit Banks. When we have accomplished these things we shall be ready to listen to anything worth while along the lines of increased production.

EDWIN W. REED, Lux, Nebraska.

WM. L. HEARRON of Palmyra, Illinois, is a pioneer worker for farmers' organizations. He was one of the three charter members of the Farmers' Equity Union.

* * *

MR. T. L. LINE, Larwill, Indiana, is National Director from Indiana and also State Organizer for that state. He is one of our most successful organizers. He has started successful Unions all around Fort Wayne, Ind. and at many other places in Indiana and Ohio, where we had been told it would be impossible to unite the farmers. When "Tom" don't organize them no one else need try.

* * *

* * * * * * * * *

* * *

Mr. V. I. Wirt of Virden, Illinois, is a good thinker and writer and as true as steel to Equity Union plan and principles. When enough farmers and wage-earners think as he does, we will revolutionize the business world in favor of the millions of workers.

* * *

* * *

MR. S. S. RAY, Bowling Green, Mo. is a National Director representing the state of Missouri. He is a member of one of our largest Unions, which has always been loyal to the National Union. The Equity Union could not have made its wonderful growth, if we had not have had the financial support of a few such Unions as Bowling Green from the very beginning.

* * *

* * *

H. O. BRATSBERG, Aberdeen, S. Dak. He is President of the Equity Creamery and Mercantile Exchange to be located at Aberdeen, S. Dak. He is State Organizer for North Dakota and was formerly president of the Reeder, N. Dak., Local Union. Give us 100 able, sincere organizers like Mr. Bratsberg, well supported by the Union and we would double our membership each year. He has done much successful work for our grand cause.

* * *

* *

MR. R. ROMER, Liberal, Kansas, is National Vice-President. His Equity Exchange never could have made the fine success it has without the assistance of Mr. Romer. He is a man of fine business ability and above reproach in character. We can always count on our National Vice-President for Equity Union.

* * *

* * * * * * * * *

* * *

MR. A. HOFFMAN, Leola, S. Dak., represents his state with great ability and credit on the National Board of Directors. He is for Equity Union first, last and all the time. He is secretary of our most successful Exchange at Leola, forty miles from Aberdeen, S. Dak. Mr. Hoffman is also secretary of the Aberdeen Equity Creamery and Mercantile Exchange.

* * *

BOARD OF DIRECTORS, MEADE, KANSAS EQUITY EXCHANGE

BOARD OF DIRECTORS, LUCAS, OHIO, EQUITY UNION.

Top row. Reading from left to right—Director J W. Keeffer, Prsident C. M. Herring, Sec.-Treas. C. W. Tucker,

Bottom row, left to right—Vice-President O. M. Ray, National

Board of Directors of Bowman, N. Dak., Equity Exchange.

Nels Boxeth, Wm. C. Farner, R. O. Bryant; S. D. Faris, O. M. Ring, Carl M. Hjerleid.